M000032384

ASTROPHYSICS
AND
CREATION

ASTROPHYSICS
AND
CREATION

*Perceiving the Universe
through Science and Participation*

ARNOLD BENZ

*Translated by Martin Knoll,
edited and revised by the author*

A Crossroad Book
The Crossroad Publishing Company
New York

The Crossroad Publishing Company
www.CrossroadPublishing.com
© 2016 by Arnold O. Benz

First paperback edition printed in 2019.

Crossroad, Herder & Herder, and the crossed C logo/colophon are registered trademarks of The Crossroad Publishing Company.

All rights reserved. No part of this book may be copied, scanned, repro-duced in any way, or stored in a retrieval system, or transmitted, in any form or by any means, electronic, mechanical, photocopying, recording,or otherwise, without the written permission of The Crossroad Publishing Company. For permission please write to rights@crossroadpublishing.com

In continuation of our 200-year tradition of independent publishing, The Crossroad Publishing Company proudly offers a variety of books with strong, original voices and diverse perspectives. The viewpoints expressed in our books are not necessarily those of The Crossroad Publishing Company, any of its imprints or of its employees, executives, owners. Although the author and publisher have made every effort to ensure that the information in this book was correct at press time, the author and publisher do not assume and hereby disclaim any liability to any party for any loss, damage, or disruption caused by errors or omissions, whether such errors or omissions result from negligence, accident, or any other cause. No claims are made or responsibility assumed for any health or other benefits.

Book design by The HK Scriptorium

Library of Congress Cataloging-in-Publication Data
available from the Library of Congress.

ISBN 978-0-8245-9937-9 (paperback)
ISBN 978-0-8245-2213-1 (cloth)

Books published by The Crossroad Publishing Company may be purchased at special quantity discount rates for classes and institutional use. For information, please e-mail sales@CrossroadPublishing.com.

For Elisabeth
In memory of the time given to us

Contents

Preface to the English Edition

For centuries the dialogue, and often the heated debate, between science and religion was dominated by the question of whether and how God might be experienced through the scientific study of nature. A recent example is our recognition of the amazing fine-tuning of the physical parameters of the universe, without which cosmic evolution could not have taken place. Some prominent authors insist that fine-tuning requires a "fine-tuner" (Alister McGrath, Rodney D. Holder) or an "intelligent designer" (Michael J. Behe), thus claiming scientific evidence for the existence of God. Professing atheists such as Richard Dawkins or Victor J. Stenger vigorously disagree, pointing out correctly that there may well be still unknown natural explanations of the fine-tuning. These atheists have reinstated what David Hume and Immanuel Kant already realized in the eighteenth century: that God cannot be evidenced by scientific methods.

On the European continent the dispute between science and religion was interrupted in the middle of the twentieth century by Karl Barth, the most influential Protestant theologian at that time. He claimed that science is irrelevant for theology and must be radically separated from theology. Barth's critics regretted that he established a barrier between the two fields that seemed to forbid any dialogue.

As an active scientist, I assume that science and religion start from different perceptions: quantitative observations and measurements on the one side, religious and existential experiences on the other. I do take seriously these religious experiences, intuitions, and visions, always aware that they are not measurable and contain a subjective element. Based on different perceptions, the methods and languages of the two also differ completely from each other. Experiments and mathematical modeling are indispensable for science; metaphoric language is necessary to express the essence of religion. Disregarding these fundamental differences has resulted in unfortunate misunderstandings.

However, I disagree strongly with Barth that dialogue between the two is futile. In a worldview shaped by science, theology is becoming incomprehensible to more and more people. The perceptions from which science and religion originate must remain distinct but should be brought into a common view and relation. They are the results of different perspectives on one reality, which at a deep, unfathomable level constitute, I believe, a unity.

In this book I explain that cosmic fine-tuning and other coincidences are no proof of God. But they are astounding, and remain unexplained. Amazement is an appropriate emotional reaction to reality. It implies that the objective world is not to be taken for granted and may well not have formed at all. Thus the cosmos is experienced as an extraordinary fluke, something quite gratuitous. In such an "aha moment" one realizes that the universe and elemental realities such as my life, our environment, and Earth itself are all undeserved gifts. However, creation is constantly taking place amid the contraries of degeneracy and death, as is palpable for instance in the course of biological evolution. We are reminded that both science and religion must face up to the sober facts of destruction and decay. Thus we need to understand why the biblical authors conclude that creation is basically "very good" (Gen. 1:31). Despite life experiences that must have been at least as negative as ours, they sensed a reality that is essentially benevolent and understood the universe to be a gift (Gen. 2:7-9). They interpreted and believed it to be a divine creation, hence meaningful and worthy of trust. Science can strengthen such faith but cannot create it.

Based on the testimony of scientists who happen to be persons of faith, some modern apologetical treatises claim blithely and abstractly that science and religion are reconcilable. This is a common view among theists, but too often it is nothing more than a postulate. Does it hold true in reality? To answer this question one must comprehend more fully the intention and scope of modern science as well as its limits. One must also reflect with more than usual care on what religion is—and is not—in connection with the faith–science dialogue. This book not only claims that science and religion are reconcilable but aims to show explicitly *how* they are compatible.

Preface

"In the beginning God created the heavens and the earth." What can this opening sentence of the Bible still tell us? Has it become incomprehensible if not irrelevant? According to modern scientific findings, the universe is three times as old as the Earth, and stars are still being formed. Astrophysics has learned much about the origin of heavenly bodies in recent years. My intention here is to describe some of these fascinating new discoveries, but also to consider their possible relation to human existence and to ancient forms of religious understanding.

The starting point is astrophysics, a science that combines observations with theoretical explanations and includes a lot of mathematics. From the viewpoint of human experience, cosmic processes and their explanations may not seem particularly relevant. But they do form the background for a fundamental understanding of the world and modern science. The new discoveries also speak to nonphysicists unable to understand formulas and details. The unimaginable expanse of the universe, the variety and refined complexity, but also the wealth of interdependence and the omnipresent cosmic network leave us stunned, amazed. The dynamics of the universe, however, have their drawbacks—the decay of all things. The decay includes our own existence and therefore calls for a reassessment of our place in the scheme of the universe. Amazement, fear, and interpretation are the three principal elements.

Creation stories such as those in Genesis do not primarily seek to explain cosmic facts, but rather to convey basic values and orientation. Genesis 1 distances itself from the Babylonian myths, which linked Earth and stars with deities. And yet a story is told in which Earth, stars, animals, and people came into being in the beginning and have not essentially changed since then. This view, however, no longer corresponds to our understanding, according to which all things in the universe have been formed in the course of time by natural means and continue to evolve. Our view of astronomy leaves open in particular the question of God's relation to the world. If creation continues to take place in our own time, then should it not be possible to see the creator at work? Admittedly, modern prayers and psalms are being written, but no new creation stories. Why speak of creation at all in the modern worldview?

This book brings into focus the differing perceptions forming the bases of science and theology and their relationship with each other. We then become aware of a reality that science cannot grasp, and we can learn how to perceive the universe as a creation.

Prologue

My decision to become an astronomer took shape in Africa. Three of us were underway in an old Fiat Topolino, south of Ouarzazate in southern Morocco, traveling in the direction of the Sahara. Hour after hour, a black asphalt track wound its way, seemingly without aim, across the undulating stony desert. It was high summer and school was out. One year from finishing high school, my two companions and I were faced with the choice of vocation. Endlessly we talked, argued, and discussed our aims and chances and our views of the world, God, and the whole business. Do we have a mandate in life? What objectives are meaningful? What's the purpose of our lives, of humanity, and of the universe? The future lay spread out before us like the southern foot-hills of the Atlas mountains, which we drove through in the glaring light of the blazing sun. The road appeared endless.

On the side of the road unadorned adobe huts occasion-ally appeared. The hostels seemed to derive from the time of the caravans. What we were offered as lodging turned out to be an empty room without beds. So we decided to spend the night under the open sky. We drove further into the uninhabited landscape until the sun disappeared behind the horizon and the night surprised us. We stopped on a rise, and each of us looked for a sandy spot between the stones to curl up in our sleeping bags.

I was lying somewhat distant from my friends. It was refreshingly cool; the oppressiveness of the heat, which reduced life during the day to a muggy type of suffering, had retreated. An unbelievable peace enveloped us. It was quiet: no din of civilization, no animals, no rustling in the air, nothing. The night opened the skies for us to reveal an unusual and overpowering splendor of the stars. The Milky Way traversed the sky from north to south. Because the air was totally clear, the stars hardly glittered and yet shone intensely. I know that one is able to count with one's eyes only a few thousand stars. The countless weak stars, which form groups and heap together into nebulous clusters, allow one to realize, without too much difficulty, that there must be a million times more of them.

The sky was alive. It no longer looked to me like the inside of a ball. The bright stars gave the appearance of being closer, while the diffuse stellar nebulae seemed farther away. Interstellar space achieved a dimension of depth. The band of stars constituting the Milky Way became separated through mysterious dark patches. These permitted the stars in front of them and beside them to glow even more splendidly. The darker the veil, the brighter the stars appeared. Everything seemed to be linked, and to constitute an impenetrable totality.

Does the unfathomable depth of the universe contain a secret that somehow relates to the secrets of my consciousness and my life? It's the big question of our existence, one that touches the core of the material and spiritual world. I noticed how the universe attracts me. The way to the Sahara led to another fascinating trip into the unexplored portions of the universe. But just as the Sahara is difficult to decipher, so is the universe.

Even before this high school experience, the idea of exploring nature through the laws of physics had begun to interest me. Until then, however, the dry reality of school courses had made me hesitate. That night in the Sahara stimulated my thirst for more knowledge and assured me, too, that this knowledge needn't stifle the sensation of amazement. With a sense of wonder, I had encountered a totally different perspective, which was not in competition with physics. On the contrary, my fascination with the quiet and mysteriously glowing stars and the prospect of pursuing new methods of scientific investigation had both captured me with their spell.

During this night in the desert, I decided to study astrophysics.

Figure 1: The Milky Way as seen with the naked eye in the southern hemisphere stretches from top left to bottom right in this picture. In the middle of it, in front of the bulk of the stars, are dozens of dark clouds. The center of the Milky Way is slightly below the middle. To the right of it the finger-like molecular clouds in Ophiuchus, at a distance of only 400 to 450 light-years from us, stretch out to the edge of the picture. (Photo: Todd Hargis)

PART I

Amazing Formation

Figure 2: The Orion Nebula at a distance of 1,350 light-years encompasses nearly the whole constellation. The picture shows a section below the three belt stars. In the center, a group of young and massive stars illuminates the gas within a radius of ten light-years. In the foreground dark molecular clouds are visible. (Photo: M. Robberto et al., NASA, ESA)

1

The Stuff We Are Made Of

1.1 Clouds in Space

On a moonless night, planets and stars are the most noticeable objects in the sky. With a bit of practice star clusters and interstellar dark clouds are easy to recognize as well, especially in the southern hemisphere. These objects are all related to one another because it is exactly in such clouds that star clusters, stars, like the Sun, and planets like our own Earth form. The story of how this was discovered is an exciting one. It had long been suspected that stars formed in interstellar clouds, but even the largest telescopes were not able to confirm this hypothesis.

Tiny dust grains hover throughout space and reduce light to a greater or lesser extent, depending on their concentration. These grains are loose aggregates of carbon and silicates less than one-thousandth of a millimeter in size. In dark clouds the concentration of dust is one million times greater than in ordinary interstellar space. Still, even in clouds, only one dust grain is found in a volume the size of a typical living room. Thus we should not envision dark clouds as dusty, unused back rooms. The so-called cleanrooms of a computer-chip manufacturer have a hundred times higher dust levels. Why then do the clouds appear so dark in the night sky? It is because the visual impression of dust grains becomes compounded over great distances, so that they completely absorb light and prevent even the weakest glimmer of starlight from making it to the outside of the cloud. As soon as the interstellar gas with its dust grains forms a cloud, the curtain is closed and the birth of stars is obscured from our view.

But not completely. Modern technology now makes it possible to observe wavelengths of light not detected by the naked eye. Dark clouds are translucent to those wavelengths larger than the size of the dust grains. During the 1960s, when it became possible to measure radio wavelengths on the scale of millimeters, astronomers were amazed to discover that signals from molecules were being emitted from these dust clouds. Until this time astronomers mainly associated molecules with planetary atmospheres, so they were at a loss to explain how molecules could form in interstellar space. Over the years it became clear that molecules not only existed in dark clouds, but also played a major role there. It turns out that the clouds are composed primarily of a gas made of molecules, while the dust makes up only about 1 percent of the mass of a cloud; hence the new descriptive name "molecular clouds." The gas is one million times denser than in the surrounding vicinity and is composed primarily of the dumbbell-shaped hydrogen molecule, which is constructed of two hydrogen atoms. By virtue of its thermal energy, this molecule moves through space at speeds of several hundred miles per hour.

While it is clear that these clouds are not simply large thunderclouds like the ones we can observe in the earthly sky, there are numerous similarities between interstellar clouds and earthly ones. Atmospheric clouds also contain gas and small particles, either frozen or liquid, which shine white when illuminated by sunlight. In contrast, on a moonless night terrestrial clouds appear dark against the starlight. Naturally, the size difference between earthly and interstellar clouds is significant—by a factor of about one quadrillion (a one followed by fifteen zeros, 10^{15}). Cosmic clouds have diameters of a few hundred light-years, meaning that even at the speed of light (186,000 miles per second) it would take a few hundred years to fly through one of these clouds. The temperature inside the cloud is colder than -300 degrees Fahrenheit, and the gas is less dense than in the best vacuum chambers of earthly laboratories. Nevertheless, enough mass exists to create thousands, if not millions, of suns like our own.

The biggest difference between terrestrial and cosmic clouds is the irregularity observed in the latter. Cosmic clouds appear in shreds and bits moving through space at supersonic speeds, producing shock waves when they collide. Densities differ, from place to place in the clouds, by several orders of magnitude. Ultraviolet radiation from neighboring stars heats the clouds and vaporizes dust at their surfaces. Burned-out, high-mass stars in the clouds explode as supernovas to form bubble-like cavities. Magnetic fields produce waves that travel from one end of the cloud to the other. But the most intriguing aspect of these clouds is that stars and planets are created within them.

1.2 Early Ideas about the Formation of Stars

In the beginning everything seemed very simple. When Isaac Newton (1643–1727) first sought to explain the formation of the Sun and other stars, he began with the concept of gravitational force. In the same way that the Earth attracts an apple until it falls from the tree, so too do the heavenly bodies and gas clouds attract each other. If the material in the vast reaches of space had originally been gaseous in nature, then chance fluctuations in density would have caused localized differences in gravity. In Newton's view, areas with slightly increased gravity could have experienced a concentration of gases leading to the formation of stars.[1] Typical of Newton and the discipline of physics after his time was the importance attributed to sequences of cause and effect. Cause is a force, in this case gravity, and its effect is acceleration. Nature was not perceived to be anarchic, but to follow a strict order that could be described by mathematical formulas. Newton's revolutionary discovery was that the rules that are valid on Earth are valid as well throughout the cosmos.

Newton's speculation on the formation of stars existed within a broader context.[2] He still took for granted that stars were permanently fixed in space. The term "fixed star" is seldom used today because it reflects an antiquated view of the universe based on the science of the time, when the unbelievably high speeds of stars were not known, and stellar movements could not be observed because of the vast distances between Earth and the stars. Although stars were immobile in Newton's view, he no longer considered them affixed to a celestial sphere but distributed throughout space. His theory was challenged by Richard Bentley, a young scholar and theologian, who asked why the stars, attracted to each other by a gravitational pull, did not collapse and combine to form a larger object. Did the law of gravity apply to the solar system but not to the distant realms of the stars? Newton held fast to his claim of the universal nature of gravity and speculated that stars were distributed equally in an infinite space so that the attraction between masses would be cancelled out. Still, Newton acknowledged that this balance in the arrangement required enormous precision. The slightest variance would eventually lead to catastrophe.

In the second half of the seventeenth century such discussions took place in a larger arena. It was precipitated by Bentley's question: if God had created such a perfect world, could he simply turn his back on it and leave it to its own devices? Challenged by Bentley's question, Newton looked for answers that would defend his theory of gravity. Characteristically he looked to nature herself for the answer. With data from star charts he was able to prove that stars in the vicinity of the Sun were actually distributed more or less equally throughout space. Yet he was not able to propose an explanation for the stability of this distribution of stars and their attraction to one another. He postulated—again in character with himself and the times—that God would intervene from time to time to prevent collapse by moving the stars back to their assigned places if they wandered astray. In Newton's worldview God not only had the task of watchmaker who created the cosmos in the beginning, but also of the much-needed repairman who kept everything running properly. Newton did not suspect God's hand directly in gravity, but in the fathomless mystery behind gravity and other forces that kept the universe running. "In Him are all things contained and moved."[3] He extended the idea of divine care from the plane of human experience to cosmic dimensions. Thus he revised the then-contemporary paradigm that the universe was a clockwork created, wound up, and set in motion by God, by presenting God as preserver of the world.

Newton's picture of God as the keeper and director of Earth met vehement criticism from the German philosopher Gottfried Wilhelm Leibniz (1646–1716), who could not understand why an almighty God would create a universe requiring maintenance. He could have made it run perfectly from the beginning. Intellectually, Leibniz belonged to an older generation whose thinking had been shaped by concepts from antiquity and the Middle Ages. In this tradition, God was seen as all-powerful and all-knowing. The concept of a distant and uninvolved God, however, was doomed to fail. Against such a backdrop, it is not surprising that modern atheism was first formulated philosophically in the second half of the seventeenth century. At this time atheism was basically synonymous with agnosticism,[4] and amounted to a supposition that God had created the universe a long time ago and didn't play a role in life. Such a God was, after all, unimportant and could be set aside without consequence.

Newton's attempt to explain the stabilization of the universe did not survive for long. The idea of a "hands-on" God lost its attraction in the early eighteenth century when gravity became a purely physical force. Yet Newton's procedure of looking to nature for answers remains the approach employed by modern science and is reflected in my own account of the origin of stars and planets.

Thanks to new methods of observation these concepts have changed markedly over the last three hundred years. Newton's basic idea that stars form through gravitational attraction within interstellar gas clouds remains valid, though his original formulation does not do justice to the currently understood complexity of the various processes. We now understand that not just gravity is important, but practically all aspects of physics, from magnetism to nuclear forces. Even the chemistry of interstellar gases plays a role. Stars in their initial formative stages are not unchanging spheres, but are characterized by dynamic processes in which a cloud core collapses, begins to rotate in a disk, differentiates chemically, and expels some material. Perhaps the most important present-day realization is that when stars and planets form, they are influenced by all prior cosmic history back to the very beginning of the universe. Science is still far from understanding completely the series of connections in stellar prehistory. At this point, let us step outside and take a journey in order to view these things more closely.

1.3 A Ramble through Our Cosmic Environment

Space beyond the solar system is not completely empty, but contains atoms, molecules, elementary particles, and fine dust particles. The differences from place to place are immense and vary by orders of magnitude in terms of density and temperature. Even the state of the material varies widely and ranges from dust and molecular gas near a temperature of absolute zero (-459 degrees Fahrenheit), atomic gas of a few thousand degrees, to ionized gas, composed mainly of electrons and protons, which is millions of degrees hot.

The solar system and closest stars are surrounded by a hot gas of twelve thousand degrees Fahrenheit. In an inch-size cube filled with this gas, there are on average only four atoms present. The gas in our vicinity is, however, about ten times denser than in the surrounding area and is referred to as "local fluff." The local fluff is our cosmic home turf. The Sun and its orbiting planets are located on the edge of the local fluff, which is 25 light-years long and between five to ten light-years wide. The width amounts to about ten thousand times the diameter of the solar system (out to the planet Neptune). The solar system flies through this relatively small interstellar cloud of hot fluff at a speed of 16 miles per second.

Space probes can detect the gas of the local fluff as it passes between the planets. The amazingly high temperatures pose no threat to interplanetary travel because the sparse density of a few atoms per cubic inch is incapable of heating up the many atoms that make up the hull of a space ship. Familiar neighboring stars such as Alpha Centauri, Vega, and Altair move through the same cloud. The Sun has plunged into the local fluff 100,000 years ago and will leave it in 20,000 years.

The local fluff floats like a downy feather in an even hotter gas of the "local bubble," where temperatures reach half a million degrees. Yet even this gas cannot harm us. If an astronaut were to open an empty one-gallon container in the local bubble, on average only 20 atoms would enter. The local bubble, with its radius of 150 light-years, must have formed tens of millions of years ago through a supernova in the vicinity of the solar system. Warm and hot zones like the local fluff and local bubble make up the greatest portion of space in our cosmic environment, the Milky Way galaxy.

In these unimaginably large spaces there are many stars, yet the interstellar gas with its high temperature and very low density cannot produce any new stars. There is, however, a third occurrence of interstellar gas in which stars can form—in the previously mentioned dark molecular clouds. These constitute only 1 percent of the volume of the Milky Way, but contain a third of the mass of all interstellar gas there. This means that molecular clouds must be quite dense. The enhanced density within these dark clouds has a surprising effect: the temperature decreases. When cold molecules collide, they begin to rotate. They are thus raised to a higher energy level and, after a while, release this energy in the form of electromagnetic waves. Molecules collide more frequently with one another in a dense gas, which radiates thus more intensely. Most of this radiation is at low energy, in the infrared wavelength range. The strength of the infrared light increases as the wavelength approaches almost one millimeter. In this way a dense gas radiates much of its heat into space and cools.

The gas in molecular clouds is composed almost exclusively of molecular hydrogen and helium. Here is a subtle problem in the realm of the molecules that has significant consequences for the clouds: hydrogen molecules and helium atoms are rotationally symmetric, thus they cannot radiate by rotation. These most common elements are therefore not visible in clouds, and they do not contribute to cooling, without which contraction to a star cannot proceed. For this reason much-less-common molecules such as carbon monoxide, carbon dioxide, water vapor, methanol, and ammonia become important.

The dust of the clouds protects the fragile molecules in these dark, cold corners of our galaxy from high-energy radiation of neighboring stars. Ultraviolet rays and X-rays

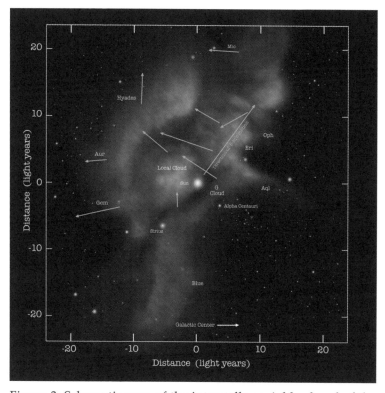

Figure 3: Schematic map of the interstellar neighborhood of the Sun. It is embedded in the local cloud ("local fluff") consisting mostly of hydrogen, indicated in blue color. The various parts move in different directions, shown by white arrows. The Sun in the center of the picture moves to the upper right (yellow). Some nearby stars such as Alpha Centauri and Sirius are indicated. The local fluff is close to the center of the "local bubble," which has a radius of 150 light-years. (Photo: NASA, Goddard, Adler Plan., University of Chicago, Wesleyan University)

from nearby stars would break molecules apart and destroy them. So without dust there would be no molecules, and without molecules no stars and planets. The interstellar dust was born in the winds of stars from earlier generations. We will deal with these winds in more detail later as we investigate the final phases of the lives of stars. The universe consists of a network of causal relationships. But how did the first star form? Stars must have formed differently in an early universe devoid of dust and molecules. This question must also be answered later.

Luckily, molecular clouds are transparent to infrared and millimeter-size waves. Our eyes cannot detect these waves because our retinas are adapted to sunlight in the visible light range. For the same reason we can also see the Sun-like stars at night. But not all members of the animal kingdom live with this same visual boundary. Some life forms, snakes for example, have eyes that can detect heat emitted from warm-blooded animals in the form of infrared radiation. When a snake looks up at the night sky, it is unable to see much of the starlight but instead sees the giant molecular clouds, some of which are about the size of a hand on an outstretched arm. For the last two decades, humans have also been able to observe this radiation thanks to continuing improvements in telescopes. We identify many common objects such as certain birds, flowers, or cars by their color. In the same way every molecule has its characteristic color or so-called spectral lines that provide information about the density, movement, chemical composition, and temperature of the gases and dust of which it is composed.

The closest molecular clouds to us are located in the Ophiuchus constellation, about 370 light-years away. Dozens of stars are forming there. The first indication that a star is forming is the presence of a spherical structure of high density, known as a cloud core. It has a diameter of about one light-year, or one quarter of the distance from the Sun to the next star. The denser a gas, the faster it can release its heat.

That is why these regions have temperatures down to -447 degrees Fahrenheit, only 13 degrees above absolute zero. At these low temperatures most molecules, except hydrogen, freeze and adhere to dust particles. The dust particles become like snowflakes surrounded by a mantle of ice, but in this case the mantle is composed of a mixture of those molecules and atoms that occur in the gas of the molecular cloud. The dust-particle surfaces act as catalysts for reactions that could normally not take place in the gas. In this way it is possible for a lone oxygen atom to combine with two hydrogen atoms to form, in thousands of years, a molecule of water (H_2O). Most of the water present on Earth today originated probably in the mantles of interstellar dust grains some 4.6 billion years ago, at a time when the solar system was forming.

1.4 The Rationality of Nature

Why don't interstellar molecular clouds collapse under their own weight as Newton had postulated? The total mass of molecules in a given cloud exceeds that of the Sun by hundreds or millions of times. If such a cloud were to collapse, no star and associated planets could form. Instead, a black hole[5] would result from which no light could escape. It would take less than one million years for a molecular cloud to collapse, which is much less than the age of our own galaxy. If clouds simply collapsed, then our galaxy would have run out of material for the creation of new stars long ago.

Newton's notion of interstellar clouds was too strongly influenced by his impression of earthly clouds. He did not know that within molecular clouds there is a constant, chaotic motion of cloud sections moving in all directions and shooting past one another. In a static cloud, as Newton imagined it, gravity would be focused toward its center. In molecular clouds, however, the attraction between cloud portions is overwhelmed by the chaos of motion. Chaos stabilizes the molecular clouds.

But not completely. The turbulence gives molecular clouds a longevity of tens of millions of years, during which time regions of increased density, the previously described cloud cores with a mass comparable to that of the Sun, begin to form. As a result of the disordered movements, the cloud becomes divided into hundreds of components. Random concentrations in density give rise to the cloud cores that later will form stars. In short, the chaotic movement of the cloud portions stabilizes the molecular cloud and permits the formation of new material in localized, dense regions of the cloud. During the lifetime of a molecular cloud no more than 10 percent of its material will be transformed into stars.[6] The radiation from the newly formed stars is so intense that the dust vaporizes, and the cloud is finally torn apart by stellar winds. The remainder of the cloud is lost in the expanse of our galaxy. The old passes away but bequeaths something new—a star cluster.

The solution to Newton's stabilization problem is found in the dynamics of the molecular clouds. A closer look reveals that the reality of the dynamics is much more complex than one might imagine. What transpires in one location of the molecular cloud has an influence on other locations. Depending on circumstances, the formation of the first star may either prevent or foster the formation of another in a particular location. Stars do not form in isolation. At birth they are intertwined with a region that is hundreds of light-years in size. Birth in clusters is simply one example of the many interrelated processes involved in star and planet formation, which we will examine further as the story unfolds. At this point in our journey, however, it is important to note that Newton's hypothesis, according to which God might be found at the limits of current human knowledge, remains unproven.

It is not clear how molecular clouds originate; they form when the density of hydrogen atoms and dust in the interstellar realm increases for some unknown reason. There are approximately six thousand giant molecular clouds in our galaxy. Within each of these, thousands to millions of stars are presently forming. In less than ten million years these clouds will disappear and new ones will form. What transpires in the molecular clouds is not completely understood; and the more we understand, the more new questions arise. In essence, the cradle of the stars remains as enigmatic today as it did centuries ago. Scientific investigations have not solved the puzzle of star formation, but have shifted the limits of our understanding. When I see bluish twinkling young stars surrounded by red hot gas within the dark entanglement of a molecular cloud, the unknown and enigmatic thrills me. Yet even more I am amazed by the variety and expedience of the processes already known to be necessary to form new stars and planets. Reasoned inquiry does not preclude wonder.

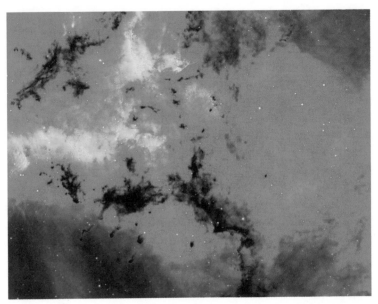

Figure 4: Dark molecular clouds in the Rosetta Nebula, three thousand light-years away, are embedded in a glowing hot gas of ten thousand degrees. (Photo: N. Wright, UCL, IPHAS collaboration)

2

When Stars and Planets Form

The previous chapter illustrated how spherical cores of gas and dust form within gigantic molecular clouds. These cloud core masses often exceed that of the Sun. Cloud cores slowly but inevitably cool and contract. Gravity increases as the gas becomes denser and the radius of the core shrinks. The force of gravity increases more rapidly than the gas pressure that opposes it. At some point gravity crosses a threshold value where gas pressure can no longer stand up to the growing force of attraction. The cloud core collapses into itself due to its own gravitational force. The collapse begins in the central portion and takes several hundred thousand years to reach the outer regions. The speed of collapse increases much as an apple's speed increases during its fall from a tree, and ultimately reaches a few tens of miles per second.

2.1 Accretion Disks

One might now assume that spherical cloud cores need only to collapse under their own gravitational force in order to form a star. Far from it! The rotational motion of an interstellar cloud is not quite zero. On the outer edge of the Milky Way the period of revolution around the center is slower than the inner portions; the cloud rotates as if it were rolling backward in its own orbit around the galactic center. Additionally, the cloud likely has maintained a certain spin derived from its origin. Cloud cores are even more unsettled, since they have gone through a wild history of formation. If a cloud core collapses, it still possesses its impulse to rotate. In the terminology of physics, it conserves its angular momentum.[1] The more the material pulls together, the faster it must rotate. A well-known example is an ice skater who, with outstretched arms, sustains a pirouette on the tips of her ice skates. If, as she spins, she pulls her arms in, she will rotate even faster because the angular momentum is maintained. A typical cloud core rotates once each million years. If it were able to collapse so as to form a star, it would rotate faster than once per second! This is not possible, as the centrifugal force at its surface would be much greater than the force of gravitation and the star would be torn apart.

So collapse does not lead to the formation of a star. Rather, the contraction ends in the formation of a rotating disk where centrifugal force and gravity cancel each other out. The diameter of the disk is typically some hundreds of times smaller than the dimension of the original cloud core or a thousand times the distance from Earth to Sun. The thickness of the disk increases toward the outer edges and remains about one-tenth the radius of the disk. The rotating structure is known as an "accretion disk" because material accumulates within it during the free fall of collapse caused by the cloud core's instability.

2.2 Why Do Planets Orbit?

The great French mathematician Pierre Simon Laplace (1749–1827) called attention to the problem of conservation of angular momentum and from this developed a theory that explained the origin of the planets. It was the first mathematically based theory about the origin of the solar system.

Laplace began with the concept of a primordial nebula, equivalent to a modern cloud core that surrounded the Sun like an atmosphere. The nebula contracted through cooling, and, because angular momentum was maintained, it rotated ever faster. For this reason the atmosphere flattened to a disk. The centrifugal force continuously increased. At the outer edge of the disk this force eventually offset the gravitational force of the Sun, so that a gas ring separated from the nebula and rotated without further contraction. After a while this process repeated itself at smaller distances until the Sun's atmosphere shrank to its present size. In the rings, material concentrated into clumps, in the way that Immanuel Kant (1724–1804), the eminent German philosopher, had suggested already fifty years prior to Laplace. According to Laplace, in each gas ring there was a dominant concentration that assimilated smaller neighboring concentrations, growing to a planet. Because the gas ring orbited the Sun, the planet had to orbit with the same speed as well. The planet attracted more gas, and the process was repeated on a smaller scale, where small bodies formed that circled the planets and merged to form moons. The leftover gas eventually remained as planetary atmospheres.

The theory of Laplace was strongly respected and gained fame during his time. It resulted in a summons to see Napoleon in his palace. In an anecdote passed down by Victor Hugo,[2] Napoleon put Laplace to the test, asking how he could explain all of creation without ever mentioning God. Laplace gave his famous answer: "Sir, I had no need of that hypothesis." Apparently Laplace assumed that he had closed all knowledge gaps in his explanation. For its time his statement had a special significance, contradicting the idea of a God who, like a watchmaker, was supposed to have created the world before the dawn of time and whose work would have left behind inexplicable gaps.

Laplace's model did, however, leave one unexplained question for his contemporaries to ponder. When an inter-

Figure 5: An image of a disk seen from the side. The light of the star—AU Microscopii, about 12 million years old and only 32 light-years from Earth—is masked for clarity. The light of the disk is emitted from tiny dust grains created by the collisions of asteroids and comets. Planets probably already exist within the disk. Such disks appear later in the planet formation and are called "debris disks." (Photo: NASA, ESA, J. E. Krist and the Hubble/ACS Science Team)

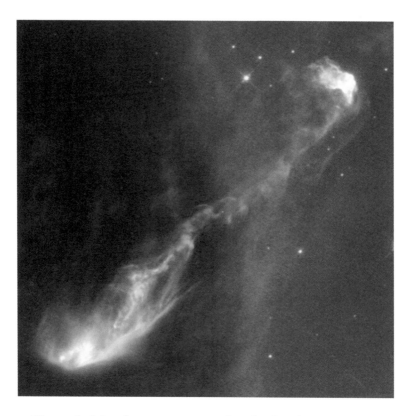

Figure 6: A jet shoots out at two hundred miles per second from the dark molecular cloud in the bottom left corner. It leaves a bright trace of shocked gas and finally hits the interstellar gas at rest outside the cloud. There, after traveling 3 million billion miles, it creates a bow shockwave known as HH-47. The jet transports angular momentum from the accretion disk, or the protostar, into the interstellar gas. (Photo: J. Morse, HST, WFPC2, NASA)

stellar cloud contracts and reaches equilibrium as a rotating disk, then it cannot continue to shrink to form a star. As later suggested by James C. Maxwell (1831–1879), it must forever orbit like a planet or a ring of Saturn. The conservation of angular momentum does explain why planets orbit the Sun. It seemed there was no reason, however, why an accretion disk should contract to form a star. Why did the angular momentum of the planets remain constant, while that of the Sun was lost? The Kant–Laplace theory could explain the formation of the planets, but not how the Sun and other stars formed from the gas of an accretion disk. This criticism was so profound and convincing that a new hypothesis surfaced near the close of the nineteenth century, one positing that a star's glancing blow to the Sun caused the planets to be hurled out.[3] Today we know that collisions between stars are very rare. In our 13-billion-year-old Milky Way galaxy, with its approximately 300 billion stars, this has occurred only with a few hundred individual stars that are like our Sun.

For a long while the process by which stars developed from accretion disks remained a mystery. Only in the 1960s did modern astrophysics begin to further develop the ideas of Kant and Laplace. In the last twenty years two real possibilities explaining how an accretion disk can lose spin and continue to shrink have been identified and partly observed.

First, though, a word is in order about the truth-value of theories formulated in the natural sciences. Has the nebular hypothesis of Kant and Laplace been proven? Certainly it has succeeded against the collision hypothesis. To speak of truth, however, would be going too far. Scientific theories are neither true nor false, but are best regarded as "good" or "bad." A good theory can explain many observations and inspire new measurements by which it can be tested. The nebular hypothesis has, in large part, withstood these tests and was further developed. The predictions of the collision hypothesis, by contrast, have not been verified. For instance, it was later found that the Sun and planets do not have the same chemical composition. The Sun is composed largely of hydrogen, while the Earth is made of iron, oxygen, and silicon. This difference may be explained through chemical separation in the accretion disk, but required a complicated addition to the collision hypothesis. Though probably not applicable to the Sun, the collision hypothesis cannot be regarded as completely false. For it was revealed that within compact star clusters, known as globular clusters, star collisions actually do play a role. Today that process is modeled with numerical simulations in computer networks. When stars are ripped apart by collisions, interesting phenomena occur, but no Earth-like planets are formed. Thus, the collision hypothesis has been definitively rejected as an explanation of how the solar system formed.

On the other hand, the original Laplace theory of a shrinking primordial nebula has long been surpassed. Cloud cores, for example, don't shrink but collapse through free fall. It was still the better of the two theories because it led in the direction of the truth. A statement may be considered true if it agrees with the facts. If a scientific theory agrees with all possible measurements, it may be taken to be the functional equivalent of truth. Measurements and observations pertaining to molecular clouds and the formation of stars will never be complete, and we will never be capable of measuring everything. We know only that portion of reality that we have observed. Therefore the scientific method leads to an adequate description of a scientifically observable reality, but never to a true theory. Theories are human constructs and can't be proven like a mathematical statement. Often theories are incomplete and compatible with only a portion of the observations. In spite of inadequate data, a majority of researchers can often consent to endorse one theory. They do so not because they have colluded in the making of some secret deal, but rather because they perceive an overwhelming force of evidence in the observed findings. One should not underestimate theories because of their provisional nature. Thanks to good theories, a space probe can fly to a comet and return with core samples.

The scenario of star and planet formation from molecular clouds is generally accepted by experts today, although the explanation has in no way reached the status of a good theory because of numerous unanswered questions. The reason for this uncertainty lies in the complexity of the processes, a taste of which is offered in the following section.

2.3 From Accretion Disk to Protostar

The answer to Laplace's problem concerning the loss of angular momentum is extremely complex, even without pursuing details of the individual processes. A large number of diverse occurrences must coincide in order for a simple stellar system to form. The complexity has its origin in this multiplicity and leads to a myriad of possible end products. The tradition of Pythagoras's music of the spheres or Kepler's elegant orbital paths of the planets sometimes lures one to imagine an easily understandable, harmonic universe. The truth is, however, much more exciting, even though ultimately unfathomable.

The physicists and natural philosophers of the eighteenth century could not know that rotating disks were accompanied by outflows of gas and dust along the axis of rotation. Such outflows were discovered in the early 1980s in the visible light range, where they appeared to be strongly bundled and were called "jets." Their speeds reach and exceed 100 miles per second. Jets extend to distances of several light-years, thereby exceeding the dimensions of the original cloud core. Where they collide with stationary gas, they push a shock wave before them, much as a ship's bow does. This shock wave shines brightly and is readily identifiable. Jets also drag a portion of the collapsing core with them. The resulting current forms a constant and mass-rich outflow of gas and dust. At least one-tenth of the gas that collapses onto young stars is lost through these outflows.

It has long been suspected that in addition to mass the jets also possess spin, or, in physics-speak, angular momentum. In 2004 observations[4] by the Hubble Space Telescope indicated that the jets possessed a rotational motion, thus suggesting an explanation for Laplace's riddle of how an accretion disk is slowed down. The angular momentum of the jets may stem from the spin of the disk. If the angular momentum of the disk is lost, then it must contract. As a result, gas in the central portion of the disk flows to the protostar. Thanks to the jets the accretion continues. The disk contracts and slowly empties while the star's mass increases.

We are now, however, confronted with the question of how jets accelerate. It is in fact almost unimaginable that dust grains and molecules can be accelerated to speeds a hundred times that of a cannonball. One explanation that comes to mind involves processes that take place in the solar corona, where magnetic forces accelerate particles. Since the magnetic force can work only on those particles that are electrically charged, we must conclude that the gas in the acceleration zone of a jet contains electrically charged particles, thus ionized atoms or molecules. How did those particles become charged? The cause is suspected to be cosmic rays that were accelerated to great speeds, close to the speed of light, by the shock waves of supernova explosions of distant stars. These high-energy protons penetrated even the densest areas of molecular clouds, collided with molecules, and then ionized them. Thus the collapse of old, burned-out stars that, as we shall see later, end in a supernova explosion contributes significantly to the formation of new stars.

The level of complexity increases once one introduces the concept of a magnetic field. For this reason most astronomers would prefer to ignore it. I would like to spare readers an explanation of how ionized molecules are accelerated by magnetic forces to form jets, since the details have not been observed and remain unexplained. Let me instead propose another way of explaining how a star-forming region can lose angular momentum. It involves its accretion disk, which rotates more quickly near the center as compared with the outer layers, in the same way that the inner planets orbit more quickly than the outer planets. One can image the disk being divided into rings, with neighboring rings rubbing against one another. Rapidly rotating inner rings transfer spin to adjacent outer rings which in turn transfer it to their next outlying neighbor and so on. This friction slows down the faster rotating rings in the interior of the disk, and angular momentum is transferred to the outer part of the disk. Magnetic fields are, however, also at work here. They couple an accretion ring to its neighbor, whereby friction is created. These extra complexities cannot be avoided.

It is likely that both mechanisms cause a loss of angular momentum. The jet may take over a part of the spin in the inner portion of the disk, while friction does the same in the outer portion. The protostar that is forming at the center obtains the majority of the collapsing material, yet the major portion of the collapsing cloud core's angular momentum is transferred to the molecular cloud. In this way the star can form as an island of low angular momentum within the stormy chaos of a molecular cloud.

By giving spin back to a molecular cloud, even small stars like the Sun contribute to the turbulence in the cloud. As we have seen, chaotic motions within clouds prevent the rapid formation of overly large stars. Star formation regulates itself. Nevertheless, the energy released by stars through winds, radiation, and shock waves ultimately destroys the fragile cloud in which they formed.

Star formation takes place in an arena that can be measured in light-years, yet is controlled by processes operating on an atomic and molecular scale. It is no surprise, then, that there are many gaps in the current knowledge about star formation. In contrast to Laplace's contemporaries, we no longer expect to find traces of God in these gaps. Rather, they provide a practically unlimited realm of research for future generations of astronomers.

2.4 Planets Form

Exoplanets, or planets that orbit distant stars, are another example of how a discovery may lead to new questions. Scientists could fare as Hercules did with the multiheaded water snake, Hydra, when each decapitation resulted in the growth of two new heads. Although the formation of planets is perhaps not unendingly complex in mathematical terms, the development of a complete theory does seem to lie in the far distance. Some portions of star and planet formation are well understood or will soon be made clearer by new measurements; others are not.

No one could have predicted that the discovery of exoplanets in 1995[5] would invalidate all previous models of planet formation. These newly discovered planets, which now number in the thousands and whose numbers increase weekly, are often giant gas spheres like Jupiter and a hundred to a thousand times more massive than Earth. Unlike Jupiter, however, they orbit close to their central star. According to earlier models of the Sun's planetary system, no giant planet could form close to the Sun, only a Moon-sized dwarf like Mercury. Did these giants originate close to their central star or did they reach this location in the course of their formation? Unfortunately, current methods of observation discover most readily those giants that lie comparatively close to their stars. Planets with the mass of Earth, as well as planets that lie at a greater distance from their star than Jupiter does from the Sun, are at the boundary of the observable and are practically unexplored.

Jupiter-like exoplanets have brought a surprising fact to light: the majority of them orbit stars with an elevated proportion of heavy elements such as carbon, oxygen, and iron. Our Sun also has higher levels of these elements than neighboring stars. These elements are apparently necessary for the formation of planets. This is not surprising when considering Earth-like planets, since Earth itself is made primarily of heavy elements. Yet for Jupiter and its relatives, hydrogen is the major element. The observations suggest that first a solid planetary nucleus of heavy elements forms which, in the case of the large planets, is then surrounded with a shell of hydrogen gas.

Accretion disks are composed precisely of that material necessary to form planets: gas, molecules, and dust. How planets form from this is, however, not clear. Certainly they could form through gravitational attraction produced by density fluctuations, which further attract material. Yet this process would take more time than is available. Dust particles that stick together initiate probably the formation process. It is likely that there are also processes at work in the gas that are self-amplifying. These are called instabilities. Perhaps even here the magnetic field plays a role.

A protostar that forms concurrently with planets makes itself noticeable in a number of ways. Its thermal radiation warms the accretion disk. Later a stellar wind ensues and ultraviolet radiation begins. Both carry the exterior regions of the disk away, until the remaining gas in the disk interior escapes. The dust concentrates in the plane of the disk and, if not assimilated by a planet, is ultimately dragged into space like a comet's tail. Accretion disks are transient phenomena. Planets have only a few million years to form within them.

There is no doubt that the science of astronomy has enabled us to learn much about planets over the past twenty years. In contrast to geographic maps of the Earth's surface, where the blank spaces of the unknown have all disappeared since the seventeenth century, the map of astronomy seems to have no firm frame. The zones of knowledge have grown larger, yet the size of the map and its blank spaces have grown even more. In terms of planetary formation, and for that matter the formation of all things from galaxies to living creatures, it is not so much that the map has a few blank spots of uncharted territory. Rather the map itself is mostly blank, strewn with a few flecks of knowledge.

2.5 From Protostar to Star

Protostars are hot gas spheres. They are heated by their own contraction, as is any gas that is compressed. Protostars radiate this energy at their surface, yet their interiors become denser and hotter over time. When certain thresholds are crossed, small atomic nuclei are fused to form larger ones. It all begins with the rare element deuterium, which fuses to form helium. Later the same happens with hydrogen, the most common element, but with greater effect. Four hydrogen atoms combine to form a single helium atom, whose mass is slightly less than that of the four original protons. The difference in mass corresponds to the energy that is discharged as heat. In these processes atomic forces are involved that release nuclear energy, and their energy reservoir is a thousand times greater than the gravitational energy that is set free during contraction. The change in the energy supply from gravitational collapse to nuclear fusion turns the protostar into a star. Its internal energy source sustains it over billions of years and assures stability.

In small stars such as the Sun, magnetic fields emanate from the interior and release their energy into the atmosphere, heating it to several million degrees and forming a corona. Even today's Sun has a corona, but that of a young star receives more than one thousand times as much energy. Rapid rotation is responsible for this. Like a bicycle dynamo, the faster it turns the more electromagnetic energy it produces. In order for a star to form, all known forces in physics must work together: gravity, the electromagnetic force, and the nuclear forces.

The wonderful interplay of all these forces also affects more distant areas around the young star. The thermal radiation of the corona is mainly emitted in the form of X-rays. This radiation penetrates deep into the disk and envelope where it knocks electrons from molecules. The X- rays leave behind traces of ionized molecules that are extremely active chemically. The freed electrons possess enough energy to ionize other molecules. The result is a network of X-ray-dominated chemical reactions. Depending on density and temperature, certain electrically neutral molecules may form that would not assemble otherwise.

Especially interesting is the water molecule (H_2O). It forms most commonly in dark cloud cores where it exists as a coating of ice on dust grains. If, during collapse, the temperature climbs above -280 degrees Fahrenheit, the water vaporizes and begins to react with other molecules. A network of chemical reactions is established in which water is continuously formed and destroyed until equilibrium is reached. The abundance of water is dictated by the ambient conditions. Stars with a mass of the Sun destroy water with X-rays to a distance about three times as great as that between the Earth and the Sun.[6] Water is not split directly into oxygen and hydrogen. Rather, it reacts with other molecules such as the chemically active, positively charged H_3^+ and HCO^+ molecules, which are also produced through X-rays. Accordingly, there should no longer be water within the inner solar system. What, then, is the origin of Earth's water? Our neighbor planet Mars also has water stored, even today, both beneath its surface and trapped in its polar ice.

There are at least three possible ways to produce water. A very common way combines oxygen and hydrogen in a catalytic reaction on surfaces of dust grains. Already in prestellar cloud cores atoms of both kinds impact dust grains where they react to form an ice mantle around the grains. Water can also form in the gas of an inner solar system when the temperature there exceeds -10 degrees Fahrenheit. A third possibility arises when an accretion disk reaches an age of about one million years, its density decreases, and X-ray and high-energy UV irradiation increases, causing ionized molecules to form which can produce water. A well-known hypothesis holds that after water was expelled from the inner solar system by the solar wind and radiation from the protostar, comets originating in the Jupiter region of the solar system finally transported water back to the inner solar system. These impacted the young Earth with a greater frequency than asteroids.

Water is an example of how molecules experience differing histories based on their distance from the protostar. If we could understand these histories and observe water and other molecules sufficiently, we might discern the physical state of accretion disks and reconstruct their history. The molecules and the ratio of their abundances are like clocks indicating the stage of development. Such time marks would help to arrange the chronology of star and planet formation and to understand the history of our own solar system. Much remains to be explained before this dream is realized, but great advances toward this end are expected in this century. We might eventually be able to read the level of development of protostars and planets in neighboring molecular clouds through the observation of molecules there. The history of molecular development might also answer the question of how common our type of planetary system is. Many diverse evolutionary paths are available for planets and stars; the possibility cannot be excluded that the solar system's path is unique. Astronomy in the twenty-first century promises to be as exciting as it was in earlier times.

Pressure in the corona of the young star overcomes the force of gravity. Thus the corona flows as a powerful stellar wind off into space and is constantly maintained through magnetic heating within the lower layers. The stellar wind and high-energy radiation of the young star wear away the accretion disk. Winds and radiation reach even farther into space and whisk away the remnants of the original cloud core. In this way a star prevents its own further growth. That seems to be why the most massive star in today's universe has only some three hundred times the mass of the Sun. Without this self-regulation, it is conceivable that the mass would continuously increase to the point where, instead of forming a star, it would cross the threshold to becoming a black hole. It is amazing how rare such cosmic mishaps are. Presumably such things happened only in the early universe. As we shall see later, the black holes at the center of galaxies may be relics of an earlier time when the self-regulation of accretion did not function in the same way as today.

Objects with less than one-tenth the mass of the Sun cannot exert enough pressure to fuse hydrogen and will therefore not become stars. The mass difference between the smallest and the largest star is only one thousand-fold. That is not much, given that the Sun's mass is two billion billion billion (2×10^{27}) tons. One would expect that a variation of a few more factors of thousand could be possible. It is amazing that star formation is regulated so that the star mass reaches the size at which hydrogen fuses at the proper rate to generate a stabile heat source permitting the formation of life.

2.6 Unfathomable Puzzles

Sometimes when I'm brushing my teeth, I try to account for all the winding pathways that the water has taken to reach my faucet. I have never been able to do so within the time decreed by the dentist for proper brushing. That illustrates the difficulty we face in understanding star formation, with its myriad of processes and unknown details. A wealth of unanswered questions remains, apparently far more than Kant and Laplace had imagined two hundred years ago. Each answer leads to new questions. A vast number of mechanisms must be operative for a star and its planetary system to form. Exploration of star and planet formation seems to be boundless.

One can fairly ask if the scientific method will ever lead to a complete explanation of things. Apparently we are apt to underestimate, over and over again, the complexity of reality. Probably the main reason we do so is owing to the circumstance that, in the controlled laboratory experiments of physics, reality is isolated from its surrounding environment. The number of interactive factors is thereby reduced, so that realities with a manageable number of variables might be tested. The example of star formation confronts us, however, with the full range of complexity born from a multitude of interactive processes. Even if we someday managed to learn all the basic equations, we would still be a very long way from understanding all of reality. Certain phenomena appear only within the context of multiple processes and variables. Every high-school teacher knows how easy it is to conduct a good conversation with a single student, but the dynamics of an entire class can take many different paths.

The present edge of science is a boundary in constant motion as we seek to advance our understanding of cosmic processes and origins. Due to the complexity of the universe, this boundary will never disappear. Yet one should not suppose that there have been no advances in the knowledge of how stars and planets form. On the contrary, today there are thousands of scientists worldwide whose research is focused solely on this problem. Every month these experts receive an issue of *Star Formation Newsletter,* which includes summaries of, and links to, the newest articles accepted by journals for publication. These proceedings amount to about 60 articles per month. Although for some of us such reports are as engaging as novelistic fiction, I often have only enough time to read titles or scan the most intriguing summaries.

The example of star and planet formation reminds me of the problem one encounters when trying to determine the length of an island coastline. This task poses no problem when a person uses a piece of string to figure proper measurement broadly on a map. Yet if one wants to reach a more precise measurement, by laying out a tape measure outdoors, the result will be larger because small coves are taken into account. The question of the coastline's length cannot be definitively resolved, because someone working with a magnifying glass could record an even greater length. Microscopes would provide a greater length yet. The question is only satisfactorily answered if we specify the scale that is important to us. If we wanted to pace the coastline for example, a yard scale would be most relevant. By the same token, we will never understand everything about star formation—but we don't have to. Perhaps one day science will understand the formation of stars to an extent that is satisfying by the standard of that time. Never, though, will we understand star formation as completely as a watchmaker understands his watch.

As recently as a few decades ago, a common opinion among astronomers was that stars were some of the simplest objects to form in the universe. Yet the more science endeavors to probe reality, the more enigmatic that reality appears to be. Enigmas are not fingerprints of a divine creator because these enigmas may, in principle, be solved. The unexplained is not compelling evidence of a hidden plan. We have yet to discover scientific grounds for refuting Laplace's claim that his concept of a directly interfering God is not necessary to explain the physical universe. On the other hand, the prophecies of past centuries and fears of many people today that one day science will explain everything are nonetheless unfounded. Science will never understand the entire universe. Laplace was mistaken about the complexity of reality. Even when a single phenomenon is explained, the entirety remains enigmatic since the explanation itself reveals new enigmas. The word "enigma" is colored by subjectivity. It remains an enigma for *us*. That enigmas exist and always will exist offends scientific reason. An enigmatic object retains a certain distance, remains unapproachable and inscrutable. This distance both astonishes and irritates. In the whole of the universe some 10,000 billion billion (10^{22}) stars have already formed. It is amazing that the most common feature in the universe is of unfathomable complexity.

So we have next to address the question of whether or not there are really insurmountable obstacles for the science of astronomy, and where the boundaries of the astronomical knowledge might lie.

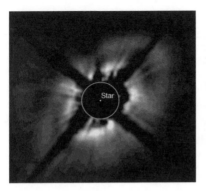 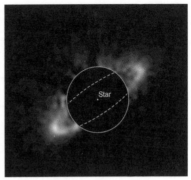

Figure 7: The stars HD 141569 (left) and HD 4796A (right) are only a few million years old and are circled by dust rings. In order to take these photos, the light of the stars was shielded with a circular disk, indicated by a white circle. The dust rings have a radius of five and two times the radius of Neptune's orbit, respectively. HD 141569 even has a second ring at twice the distance. The rings are probably leftovers of an accretion disk that was carved out by a planet. The dust grains are most likely still coated by water ice. (Photo: A. Weinberger [UCLA], B. Smith [UH] et al., NASA, and ESA)

3

Boundless?

The current state of knowledge in astrophysics holds that the universe developed with a tremendous dynamism. The formation of stars and planets involves processes that build upon cosmic events from an earlier phase of the universe's history, such as the formation of molecular clouds and galaxies. A fundamental characteristic of the universe is its continual development through processes that rise to ever-new levels. Time plays a more important role in this development than was once thought. What exists today did not form in a mythical past before the beginning of time but rather with time, and as a consequence of time.

Time produces changes and new developments, and since 13.8 billion years ago has proceeded from cause to effect. The "Big Bang" scenario[1] is generally accepted among scientists, even if the uncertainty of theories relating to events taking place in the early universe increases as one approaches the absolute beginning of time. Not a single object found in today's universe was formed at the beginning of time. Even ordinary matter did not form at the very beginning, since its components (protons and neutrons) only originated a millionth of a second later. This may seem to be very close to zero, yet physical conditions changed immensely during this tiny interval. The formation of matter proceeded in the framework of space and time subject to the laws of physics, like those that are familiar to us from laboratory experiments. But what happened within the first millionth of a second is less well understood or remains largely unknown.

The cosmos did not begin the way a theater production does, where actors and scenery are at the ready, the curtain opens, and the play begins. In the modern worldview, the course of cosmic development is much more dramatic. If aligned with the theater metaphor, it would include the construction of the building, the installation of a workshop for stage design, and a school of acting. Everything would then collapse, be built up again, burn and disperse, and continue on this cycle but slowly evolve until finally our story is played out. The universe is not a piece of clockwork that simply ticks away after its start. It develops continuously, always opening into new dimensions. Its development began with the formation of elementary particles in the nuclear realm, followed by atoms and galaxies, molecules, stars, planets, and finally life forms and human society. Processes in the universe are immeasurably diverse and surprisingly creative.

3.1 The First Stars

Star formation in the universe also has a history. Thus, we saw in the previous chapter how heavy elements such as carbon and oxygen play an important role in modern processes of star formation. These elements could only form, however, in earlier stars. What came first, the chicken or the egg? As is the case with the poultry, the answer is that earlier stars were different and formed differently.

The light of distant galaxies reaches us only after millions or billions of years. Thus there is the opportunity to look into the past. Starlight shows us stars as they were long ago. Many no longer exist today. The farther away an object is, the farther into the past the journey goes. Such gazing into the past reveals that the heavy elements increased in abundance through the course of cosmic history. The older the galaxy, the stronger the spectral lines of these elements appear in the starlight. This observation confirms the hypothesis that heavy elements form in stars and were not present at the very beginning.

Big stars are different from the Sun. To begin with, they don't live as long, even though they contain more fuel. In the interior of massive stars the pressure is greater, so that the fusion of hydrogen occurs much faster. A large star is wasteful with its supplies. It also has a strong stellar wind, through which it loses more than half of its mass during its short life. In this wind, light-absorbing carbon, oxygen, and other heavy elements absorb the starlight and are carried away by its momentum as if they were sails. Heavy elements are the combustion by-products of the star's own energy supply. If a star has a mass ten times that of the Sun, it lives only some ten million years, or a hundred times less long than the Sun. Because there were previously many massive, short-lived stars, their ashes quickly accumulated in the early universe. The majority of the oxygen that we breathe today originated in the early universe from massive stars that have long since burned out.

The very first stars had to form without heavy elements, without magnetic fields, and without cloud turbulence. They were not just larger, but completely different from today's stars. Their predecessor clouds had temperatures of about -90 degrees Fahrenheit. This is significantly warmer than today's molecular clouds, which have temperatures of -350 degrees. The warmer clouds of the early universe fragmented less quickly. In their interiors a single cloud core formed with the mass of a few hundred Suns. The cloud core collapsed on a central object, because the precursor clouds possessed little angular momentum. Due to the absence of heavy elements needed to power a stellar wind, accretion ceased only after the internal nuclear oven had been burning long enough to produce them. There are clues suggesting that the first stars were hundreds of times more massive than the Sun.

Before the first stars there was no light other than the infrared range thermal radiation left from the Big Bang. Then the first stars pierced the darkness, each with a luminosity of about ten million Suns. These stars had no planets, and their surfaces had temperatures of around 200,000 degrees. The Sun, with its 10,400 degrees, radiates white light. At higher temperatures maximum radiation is shifted to the shorter wavelengths of blue light and then to ultraviolet. The monster stars of the early days glowed in the extreme ultraviolet range. Without Earth's protective atmosphere, such rays would be deadly for humans because they damage molecules of the skin and cause cancer. Higher energy rays alter, that is, ionize, atoms by stripping them of electrons. The radiation of the ancient stars was so strong that it ionized all the intergalactic gas. Hydrogen, the most common atom found in gases between galaxies, was split into its component electron and proton. This is still the state of hydrogen in intergalactic space today. The first stars changed the universe forever.

The earliest stars have long since disappeared. Because of their gigantic mass, the life span of such objects was only about one million years. When the fuel in a star's interior is fused to helium, it burns to carbon, then to oxygen, and in massive stars on to iron. Ultimately these energy reserves are used as well and the star collapses on itself. During the implosion new energy sources are produced. Even heavier elements such as gold and uranium are created. Thanks to this energy the outer portions of the stars are flung off. The hot gas shoots off into space and glows so brightly for a few weeks that it may be seen in the entire universe. The phenomenon is known as a supernova when it occurs in today's stars. Among the extremely massive earliest stars such a process would more properly be termed a meganova. Although none have been discovered, their traces are not to be overlooked, since intergalactic gas is full of their ashes, including gold, silver, titanium, and uranium. The remaining material in the center of an early star then became a black hole. Due to the massiveness of the stars and because the pre-supernova stellar winds were minimal and mass losses were relatively small, the black holes that formed in the early universe must have been tremendously large. Some astronomers suspect that they have gathered together at the centers of today's galaxies. The mass of a black hole at the center of a galaxy may exceed today that of a billion Suns.

The first giant stars appeared about 150 million years after the Big Bang. Their formation reached its acme some 250 million years later, and by 600 million years after the Big Bang the time of the earliest stars was over. They represent a relatively small episode in the history of the universe. It was the first and greatest fireworks display of all times. Before their time extends a dark epoch from which no radiation has yet been discovered. It is a time known as the "Dark Ages," by analogy with Europe's early Middle Ages.

The universe did not, however, begin in darkness. A glaring light radiated by the hot gas dominated all other energies. Since before the Dark Ages the universe was so dense and hot that its atoms had not yet been surrounded by electrons, light was not able to propagate. The universe was an opaque, completely ionized gas. When it became transparent 380,000 years after the Big Bang, it glowed in red light like every other object having a temperature of 5,500 degrees. With the expansion of the universe the color shifted from red to infrared and ultimately toward even longer wavelengths. This radiation is detected today at a wavelength of some millimeters and appears as a background to all other heavenly bodies that are observed at this wavelength. This shimmer from the phase of developing cosmic transparency is the earliest light that can be detected.

How sure are we of all this? I am occasionally asked that question within my circle of friends, although I'm probably confronted more often with unspoken doubts. In truth, we are not equally certain about all explanations and theories. Doubts are appropriate. I view the scenarios presented in this section as robust as the knowledge of the Moon was fifty years ago. When the astronauts finally landed, they discovered much that was new. And yet, as calculated in advance, their capsule did not sink into the dust and they did not burn their feet on the Moon's surface. The early universe is a field of research that will develop rapidly in the next decades thanks to new instrumentation.

3.2 The Big Bang: Our Horizon in Time

Following clues in today's cosmic gas, we can travel even farther back in time. Most of the universe's helium originated a few minutes after the Big Bang, when the universe's temperature was one billion degrees. Helium originated from protons, the nuclei of hydrogen, and neutrons that collided and fused to form deuterium. The latter collided with another proton and neutron to form a helium nucleus. A few of the helium nuclei reacted further to form lithium, the next heavier element on the periodic table. In principle, this process would have continued given enough time. This, however, was not the case due to the expansion of the universe so that the high temperatures and densities needed for nuclear synthesis only lasted for some five minutes. The observed proportion of helium in the universe is one quarter of the total mass, which places strict limits on the physical parameters and development of the universe at that time. It conforms precisely to today's cosmic standard model, as derived from background radiation.

There is good reason to assume that previously unknown elementary particles from an even earlier phase of the universe will be discovered in the near future. They have already announced their presence in the universe through their gravitational force and are referred to as dark matter. It exceeds ordinary matter by a factor of five. Dark matter thus dominates all other forms of mass and was important to the origin of the first giant stars and galaxies. When the solar system, on its 140 miles per second orbit around the Milky Way's center, moves through this sea of dark particles our bodies are pierced by several millions of particles per second. We feel none of this. During our lives we take no notice of dark matter because it has no, or only extremely weak, interaction with itself and with ordinary matter. It radiates no energy, is therefore invisible and does not cool off. In the formation of today's stars it no longer has any influence.

Perhaps someday we will detect the distant rumblings of the Big Bang in gravitational waves from the first moments of the universe. Gravitational waves originate when large masses are accelerated. Such waves could have formed during the inflation phase, which took place some 10^{-35} seconds after the Big Bang and during which time the universe expanded extremely rapidly for less than 10^{-32} seconds. If detected, they would likely represent the earliest signals of the universe.

Since stars are still forming today, a boundary in time must exist, because stars bind and use hydrogen. As this process cannot proceed in perpetuity, there must be a beginning and an end. What is known about cosmic expansion and background radiation enables astronomers to set the age of the universe at 13.8 billion years. The value, which has been determined by various methods, is claimed to have an astoundingly small margin of error, even though most information about the origin of the universe remains unclear.

In terms of physics, the origin of the universe may be envisioned as a spontaneous fluctuation of a vacuum. It is speculated that during this process the laws of physics, such as conservation of energy and charge conservation, applied. Quantum theory permits variations within the range of uncertainty; the energy at a given location in space fluctuates in time. At time zero of the Big Bang such a fluctuation may have stabilized, perhaps enabling a yet unknown quantized form of gravity to play a role. The universe corroborates these assumptions, inasmuch as they can be tested. The energy of gravity is negative, because it takes energy to separate two bodies attracted to each other. The positive energy of matter and its expansion more or less cancels out the negative energy of gravity. Because the physics in play under conditions of the universe's earliest phase remains unknown, very little is known about the first nanoseconds. We must especially wonder if laws of nature existed at the very beginning, and if they were the same laws we know today. And was there a "before"? The answer lacks a concept of time, since an integral part of modern physics involves time that is defined through measurement. Observations of the universe before the Big Bang are likely to be not only forever impossible, but are also unimaginable.

Time expresses itself in that a cause precedes its effect, and therefore appears in the realm of physics wherever changes take place. This is seen in its simplest terms in Newton's second law, which dictates that momentum (mass times velocity) changes by an amount increasing with duration. The factor controlling the proportion between momentum change and time interval is what is called force. The change in momentum increases the longer that the force is at work. Said another way, time permits force to develop its potential. In physics time is defined through a measurable change caused by a particular force. Today, a second of time is defined by a certain number of oscillations of the cesium 133 atom, meaning that it is defined by a process that is periodic. If changes, fluctuations, or periodicities cannot be measured, time in terms of physics is not defined. This applies to the universe at the zero-point in time and before. By this logic one can say nothing in terms of physics about a "before" the Big Bang. It is an insuperable barrier of physics.

The Big Bang is an especially interesting field of study for mathematical physicists who advance and test new theories of matter, space, and time regarding this primordial event. The Big Bang also possesses a certain mythical appeal that serves to enliven such research. From a purely physical standpoint, there is no essential difference between the origin of the universe and the origin of stars. It is the same type of physics (ultimately gravity and quantum field theory) that is applied to characterize both processes. This is an important point. If we want to explain the origin of the new in the universe through physics, then we have no other physics at our disposal other than that which we know from the laboratory. It is amazing that it is sufficient to let us explain, to some extent, most observed processes in the universe.

It will never be possible to observe the actual zero-point in time. Therefore it is important to exercise caution. Regarding the subject of the origin of the Big Bang, one commonly encounters today a kind of speculation about nature that, while mathematically correct, cannot be substantiated through observation. This hypothetical cosmology is therefore not subject to the successful discourse between theory and observation that has existed since Galileo's time. Theories are relevant to physics only if they include observable evidence. If we are to assume a measurable reality, then time and the interplay of cause and effect have begun just at the zero-point. Nevertheless, the application of theories such as quantum mechanics is plausible in the early universe, and attempts exist to explain universe formation. This demonstrates that the origin of the universe may be considered a natural event, even if we can't understand it.

After giving public lectures, I am often asked what came before the Big Bang. The question aims at the transition between the nonexistence and existence of the universe. Is Laplace's claim that the God Hypothesis is unnecessary still relevant at this point? As with Laplace's theory of the origin of the planets, we in our quantum speculation of the origin of the universe have not invoked the concept of God. It is astounding that many contemporaries look for God in the Big Bang of all places. It may be argued that instead of unprovable speculations about the workings of physics under the extreme conditions of the zero-point the term "God" may be introduced. This will not be appreciated by most scientists who have learned to never call an unsolved problem "God." A divine intervention at the Big Bang could never be proven scientifically.

I venture to claim that a religious explanation of the Big Bang is a precarious path also in theology. Such an explanation may stem from the remains of mythological visions of a creation before time, from the philosophical concept of a prime mover, or from the wish to place God somewhere in the physical universe. The Big Bang would then separate the realm of the Divine before from the physical realm after, reminiscent of the iconostasis in Orthodox churches, which separates the profane portion of the church with its believers from the holy part with the altar where only priests are allowed. There will be more to say later about the dividing wall between the world and God. Theologically it doesn't make sense to constrain this division to an event that today can only be observed in a limited way at the edge of the universe at a distance of billions of light-years.

Those of us spoiled by scientific advances might find it difficult to leave open questions pertaining to the Big Bang. Here I find myself of the same mind as the famous American physicist Richard Feynman, who said: "I can live with doubt, and uncertainty, and not knowing. I think it's much more interesting to live not knowing than to have answers which might be wrong."[2] The Big Bang is like a horizon in time, a boundary beyond which we will likely never be able to look.

3.3 Black Holes: Horizons in Space

With black holes a second boundary appears, this time in space. The interior of a collapsing star that does not get ejected as supernova material cannot be supported by gas pressure and continues to collapse. Gravity increases until the stellar remains vanish behind what is known as an event horizon. Within this sphere-shaped boundary, space is bent. In other words, no rocket, not even light, moving with the greatest speed can escape a black hole. Even light moving directly upward from the center of a black hole would fall back, as a stone thrown upward. Black holes can be detected only indirectly by their gravitational effect on neighboring stars. The event horizon separates normal space in which we live from the interior of a black hole. What the interior looks like we don't know and will never know. There could be a void in the center containing an Earth-like planet. Gravity at the center would cancel out, and at the surface of the planet it could be the same as on Earth. Maybe there is a paradise world there or a hell full of dragons. The inhabitants would not be able to send any signals about their existence out into space. They would be forever separated from the rest of the universe by the event horizon.

Of course this is all fantasy! But the event horizon is a real boundary in space that makes it impossible to view or study the area behind it. The edges of black holes are similar in many ways to the time horizon of the Big Bang. Science is equally at a loss when confronted by the question of what lies beyond. Because we can't peer past the horizon, we also can't make any scientific statements about what lies beyond. Without doubt there will be advances in the understanding of the border region outside of black holes (for example, by way of a quantized gravitational theory). The boundaries of human knowledge move and will never be set in one place, yet we must likely get used to the event horizon.

In our own galaxy black holes occur in two size classes. The smaller black holes are derived from stars that have masses exceeding approximately twenty times that of the Sun. After these stars have used up their nuclear energy, they collapse and eject a supernova shell. If what remains is sufficiently large, it will contract to form a black hole. The orbital speeds of stars that orbit black holes, which are often in multiple stellar systems, reveal the mass of the central object. From such observations about one dozen good candidates for black holes are known. Black holes in the second size class, larger by many factors of ten, dominate the most central region of galaxies. In the central part of the Milky Way, for example, there is a supermassive black hole with a mass four million times that of the Sun.

To search for God in the realm prior to the Big Bang seems pointless. Why and how should we do this? It would be even more absurd to assume God's domicile in black holes. The question of God's presence is, however, too interesting to be made taboo. Yet the concept of God does reappear at a very different boundary of astronomy, one that will be considered in the next section.

3.4 Silence of the Stars: Limits of Methodology

Sometimes I observe stars in a way quite distinct from that which utilizes high-tech instruments and in a way that does not seek to understand them in a scientific sense. On a clear night in the mountains or in the desert the starry heavens are simply overwhelming. The American poet Walt Whitman (1819–1892) described this alternative way of observing stars in the following poem.[3]

> When I heard the learn'd astronomer;
> When the proofs, the figures, were ranged in columns before me;
> When I was shown the charts and the diagrams, to add,
>   divide, and measure them;

When I, sitting, heard the astronomer, where he lectured with
*   much applause in the lecture-room,*
How soon, unaccountable, I became tired and sick;
Till rising and gliding out, I wander'd off by myself,
In the mystical moist night-air, and from time to time,
Look'd up in perfect silence at the stars.

Here Whitman refers to two kinds of human experience regarding stars: first the objective, scientific observations and measurements of the astronomer and then the poetic, transcendental, or mystical experience. The latter kind of observation does not permit a person to remain in a passive role. Instead it requires the person himself or herself to become the instrument of observation. Whitman was directly involved in this second type of observation of the stars. He was personally affected by it, and, figuratively speaking, he came into resonance with the universe.

The first time I read the poem I was disappointed at the apparent disdain Whitman showed toward the science of stars. There is, however, an internal relationship between both parts of the poem. The events in the poem don't take place on just any dark night of the sort so common in electricity-free America of the nineteenth century, but on the night of the astronomer's presentation. I understand the poem to mean that the knowledge of the astronomer opened the way toward the poet's human-grounded, personal experience of the stars. Astronomy confronted the poet with a new worldview and opened for him a new horizon within which, painfully and rationally, he had to find himself again. Amazingly, and against the backdrop of visible reality, Whitman was able to do this, when finally he experienced unity with the cosmos. The astronomer's presentation brought the sky closer to him in a rational way through scientific facts and explanations. That is why the ordinary scenery of stars thereafter spoke to him on an emotional level. He personally takes part in a mystical experience that he describes as "silence."

Even as a professional astronomer, duty-bound to conduct objective science, I have experienced moments as described by Whitman. Some readers may recall similar experiences of their own, others may have forgotten them; and only a very few have written poems about them. They are unforgettable moments in which time seems to stand still. They may be life's milestones where all becomes tranquil or where everything changes. Thus they have a concrete and real effect, and must be considered as part of the reality in our life.

We should forgive Whitman, a person of the century of romanticism, for placing more worth on the second experience than on the first. Even in our twenty-first century, and especially because the astronomical knowledge has increased tremendously in the meantime, it is important to connect both types of perception. When emotion meets reason, a direct encounter with the universe is possible in the way that Whitman so vividly described. This conjunction suggests that the sphere of human experience is larger than the realm of science. The perception of "silence" is not a scientific observation. The silence of the stars cannot be explained through astronomy, and shouldn't have to be. It is not a part of astronomy and lies beyond the boundary of science.

The methodology of astrophysics has led us to recognize strict boundaries in space and time. The initial choice of measurements and observations that this methodology requires, however, sets a limit from the very outset. It is a less obvious boundary, because it is only discernible from a perspective outside the sciences. To discern this "outer boundary" is vital for the dialog between science and the humanities or theology. Without it, the topic of the next chapter would be incomprehensible.

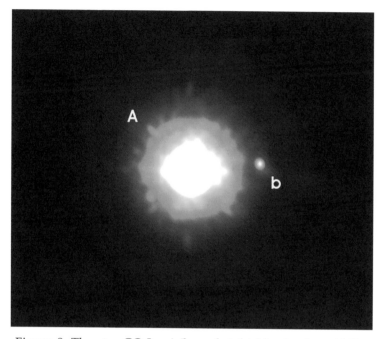

Figure 8: The star GQ Lupi (large bright blur in the middle, marked A) is only a million years old. It is accompanied by a second, dimmer object (marked b) of the same age having a mass only slightly larger than Jupiter. Object b was possibly the first planet outside the solar system directly imaged. The planet is at a hundred times the Sun-Earth distance from its central star. The image shows its heat radiation. The young planet radiates strongly as it is still more than 2,700 degrees and has not yet completely cooled from its formation. (Photo: R. Neuhäuser, ESO)

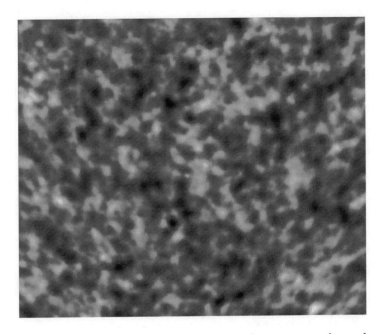

Figure 9: Waves at millimeter wavelength propagate through the entire universe. If all emission sources from the universe's foreground are eliminated, a background remains. It originated 380,000 years after the Big Bang, when the universe became transparent. The picture shows variations in intensity of this radiation in an area equivalent to an outstretched hand. The size of these fluctuations allows inferring the energy density of the universe. (Photo: S. Masi et al., New Astronomy Reviews, 2007, Boomerang Project)

Figure 10: Hubble image of the Milky way's center in the infra-
red light, which penetrates the gas clouds in the foreground.
The black hole in the center is surrounded by stars and dust,
but is not visible. Its mass is 4.3 million times the solar mass.
The size of the image is 60 light-years. (Photo: NASA, ESA, &
D. Q. Wang)

4

Origin and Creation

On a bright Monday morning at my Institute in Zurich a doctoral student burst into our regular coffee group gathering and exclaimed, "Did you know that 40 percent of scientists believe that God made people?" After an awkward silence I asked him how exactly such people envisioned the creation of humans. He evaded this question, emphasizing instead his own belief that humans had evolved from apes. I still wanted to know to what sort of creation he thought the other 40 percent of scientists affirmed. His answer was simple, "Either God made the first humans, or man and ape evolved from a common ancestor." For a few in the group, this dichotomy sounded terribly naïve, yet none could or would counter him with a convincing argument why God couldn't create through evolution. I was taken aback to find that, because mentioning "creation" has become something of a taboo, even among academics, the term has by now taken on a popular meaning close to that held by fundamentalists who imagine that the first human was literally made from dirt. Surprisingly, though, one encounters this term in the media quite often. How is one to understand "creation"?

This chapter deals with the contrast between scientific knowledge regarding star and planet formation and theological concepts of creation. I contend that both are able to be understood and framed in a productive relationship *if they are recognized to originate from separate perceptions.* The two ideas converge only in persons who allow themselves to be amazed.

4.1 Physico-Theology

Creation stories of various religions relate how God or the gods created the stars and planets. It is worth noting that in these narratives, including the two found in Genesis 1 and Genesis 2–4 of the Bible, creation takes place in a mythical time. Earth, Moon, stars, planets, animals, and humans came into being during a time that has long passed and is no longer approachable. The world produced by such ancient creation events is static. If any change takes place, according to such creation stories, it is for the worse (for example, the Fall of Man, or the shift from the golden age to the iron age in Greek mythology). Already Greek philosophers of pre-Socratic times were not satisfied by myths to explain the world's origins.

In the seventeenth century, shortly before the natural evolution of the Earth was discovered in geology, religious explanations of natural phenomena reached their acme. It was a time of major discoveries in biology and geography. Physico-theologists[1] (Greek *physis* = nature, world; note difference from modern term "physics") were encyclopedic in collecting marvelous details of knowledge, and never tired of explaining how God in his omnipotence, wisdom, and goodness had clearly arranged everything. More pointedly, God in his grace made the frog green so that the stork would not find him in the grass. Nature was consequently thought to offer reasonable proof of the existence of God.

David Hume (1711–1776), and later Immanuel Kant, unmasked the physico-theological proof of God to be circular reasoning. This is the false logic: "God is assumed to be wise; we observe that stars are functional; therefore they were created by God; therefore God exists." The assumption that "God is wise" implies his existence in advance, and therefore the argument is not substantiated. The conclusion is also invalid, because functionality may have other causes. The sciences, especially biology, have subsequently demonstrated this. With the establishment of an evolving world in which all things formed over the course of time, physico-theology found itself in the role of competitor with the sciences. Physico-theology also received criticism from theologians, who insisted that God should not be imaged as a designer working at a drawing board. Rather, as Moses discovered at the burning bush, God revealed himself at the exodus of the Israelites from Egypt as "I am who I am."[2] The biblical God does not create like an engineer, but rather so as to confer free will and responsibility on his subjects. If one wished to find evidence for God in the world, one may also look to suffering and catastrophes, in which case it could be concluded that God is a bungler.[3] Karl Barth (1886–1968) strongly rejected the ideas of natural theology, into which physico-theology had evolved, by stating that "there can be no scientific . . . aids in relation to what Holy Scripture and the Christian Church understand by the divine work of creation."[4]

In the seventeenth century, physico-theology undoubtedly inspired an interest in nature. Scholars developed and improved instruments, like the microscope of Jan Swammerdam (1637–1680), to observe creation more exactly. They learned to direct scientific questions toward nature. The astounding results provided by this new type of research gave impulse to new theological questions and interpretations. Therefore physico-theology became popular as a form of devotional literature, especially in Protestant circles. There was, however, a negative side: in the attempt to assure the existence of God through scientific observation, the physico-theologists reduced the concept of God to the scientific plane of observation and explanation. In the words of Whitman, this is the plane of numbers, columns, tables, and diagrams. It is arguable that God might appear in this plane, presumably somewhere within the gaps of our knowledge. Yet these gaps, which were explained at the time by divine interventions, belong to the same plane as that of science and are therefore susceptible to study by scientific means. As Laplace concluded one hundred years later for his astronomical explanations, God as an explanation for natural facts in the sciences is not necessary. Modern physico-theological constructs with snappy names such as Creationism and Intelligent Design fail in the same way.[5] They retain the idea of God as a pseudo-scientific explanation for nature. Yet our recognizing such pitfalls need not prevent us from marveling over the amazing processes operative throughout the universe, a disposition that does tie us in some measure to the physico-theologists.

4.2 Why Speak of Creation?

After this excursion into the past—a past that touches the present—the question must be asked if we should avoid the concept of creation and limit ourselves at best to a concept of God defined by existential life experiences. Clearly the idea of creation before time is not tenable in a universe where all things formed after the Big Bang and where new things continue to form. Previous chapters demonstrated that God is not a necessary component of scientific theories explaining the origin of stars and planets. So why even discuss God in relation to nature?

The oldest writings in the Bible that deal with creation are the psalms. Their present versions date from around 200 BCE. Individual parts, however, have their origins in the time of King Solomon, whose rule lasted from about 965 BCE to 926 BCE.[6] These parts are certainly older than the first creation text (Genesis 1–2, 4) in the beginning of the Bible and perhaps older than the oldest parts of the second creation story. Psalms retain ancient verses of liturgical chants. These are for the most part dirges or songs of thanks, either from individuals or as expressions of collective history. There are also hymns (especially Psalms 8, 19, and 104) that through subtle poetry chronicle God's work in nature. Why do psalms speak of God in nature? The experiences of the Israelites with Yahweh occurred primarily in their own history and not in nature. The psalms apparently are not concerned with proving the existence of God or explaining the universe. Rather, they present a concept of creation that emphasizes similarities in structures and processes between the natural world and personal or historical events.

In the psalms, creation is praised. One kind of experience that substantiates this impulse even in modern times is wonder. Humans did not create nature themselves, yet the living conditions on Earth are amazingly congenial compared to those one could imagine prevailing on other planets and throughout the universe. Certainly life had billions of years to adjust. Psalmists point out the excess of goodness that exists beyond mere survival. This experience has both a personal and a subjective element: the world is not purely good; but from the perspective of the psalmists, it is ingenious and worthy of praise. The psalmists present no statistics or proof that the average of good and bad is slightly tipped toward the positive. Neither do they argue that there is good to be found even within all bad. They allow themselves to be directly spoken to by the sun, by rivers, animals, and grass.[7] Yet it is not an ideal world, because there are other experiences as well. If we investigate more deeply, we see that the psalmists' sense of wonder has a prehistory, which involves terrifying experiences such as flight, wars, deportation, loss of statehood, fears about the future, and concerns about survival. Among the psalmists a sense of wonder held the day.

To speak of creation (or a creator) is appropriate only if such a discourse ultimately pertains to concrete perceptions. It is, however, a very particular type of perception—not one built from scientific measurements that someone could glean from technical literature. One must observe for oneself and become convinced that everyday things are good within the context of an endangered and vulnerable world.[8] Allowing a sense of wonder, an individual not only perceives the world in an objective and scientific manner, but reacts to it with a certain feeling. The term "creation" must not involve explanations of scientific measurement or observation. Creation has something to do with reality we perceive on a plane distinct from that examined by science.

4.3 The Parable of the Invisible Gardener

The British philosopher Anthony Flew criticized the natural-theology concept of God in his satirical parable of the Invisible Gardener:[9]

> Once upon a time two explorers came upon a clearing in the jungle. In the clearing were growing many flowers and many weeds. One explorer says, "Some gardener must tend this plot." The other disagrees, "There is no gardener." So they pitch their tents and set a watch. No gardener is ever seen. "But perhaps he is an invisible gardener." So they set up a barbed-wire fence. They electrify it. They patrol with bloodhounds. But no shrieks ever suggest that some intruder has received a shock. No movements of the wire ever betray an invisible climber. The bloodhounds never give cry. Yet still the Believer is not convinced. "But there is a gardener, invisible, intangible, insensible to electric shocks, a gardener who has no scent and makes no sound, a gardener who comes secretly to look after the garden which he loves." At last the Skeptic despairs, "But what remains of your original assertion? Just how

does what you call an invisible, intangible, eternally elusive gardener differ from an imaginary gardener or even from no gardener at all?"

The belief of the natural theologian remains untouched by hard reality, yet loses its relationship to perception and thus, as Flew suspects, becomes futile. The question raised in the parable is about the experiential basis of faith. Flew must accept, however, that his vision of God is today open to theological criticism. Even though this vision of God originated in the Christian West, it has little in common with biblical notions. The investigators in the parable looked for God in the wrong place. Their story, for example, could continue in this manner:

Because they were so absorbed in experiments and analyses, and also because of their familiarity with the place, these researchers were no longer able to see the beauty of the garden. The day of leaving, the Skeptic wandered in a reflective mood through the garden and found himself standing unexpectedly before a magnificently blooming red rose. It stood large and alone in a meadow. The Skeptic was captivated by the luminous color, the delicate form of the petals, and their contrast to the thorny stalk. The flower reminded him of something long forgotten. It warmed his heart, and he felt an inner connection with the plant. The thought struck him that it was part of a whole that included not just the garden, but him as well, and that in the end he, too, was part of an all-encompassing beauty. He went on to ask himself if his perception was self-delusory. Is beauty just an illusion, a trick of synapses in the brain? Yet he felt something undeniable, a sense of happiness that continued to resonate within. Later, as he left the garden, even his colleague noticed the change in him. "We have investigated everything except the beauty of the flowers," said the Skeptic. The other answers:

"Beauty is not measurable or provable. You experience it when you let it speak to you. Beauty is neither an assumption nor a statement, but rather an overwhelming experience. We should have known that it is the same with beauty's creator, who is only recognizable if we, full of wonder, allow ourselves to be embraced with his goodness. We see him only if we are personally ready for it and thus contribute our part to this perception. Surely he was in the garden, but we were too busy with our measurements to perceive him."

To sum up: Today the term "creation" is often used to describe prescientific attempts to explain natural phenomena or as a synonym for the natural formation of material objects in science. Here creation means something different from the question of what processes were causally involved in the development of the cosmos. The term, as used in the Bible, was originally based on personal and historic experience. *Like beauty, creation is therefore experienced on a plane different from that of science.* What are these experiences? Perceptions that lead to a theological notion of creation are addressed in the following chapters. Toward this end, we must first expand our cosmic view in the direction of the future. If the formation of the new elicits a sense of fascination and wonder, continued development shows a more sinister counterpart—that of decay. The latter deserves equal treatment in the history of development and must therefore find a place in any realistic notion of creation.

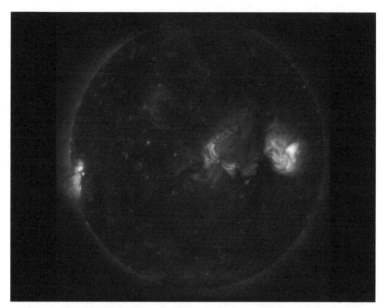

Figure 11: The outer atmosphere of the Sun, the corona, is at a temperature of more than a million degrees and glows in X-rays. They are made visible in the picture by bright color. Bright regions are located above sunspots and are hotter and denser. The hot gas has excess pressure. In dark parts it escapes into space as the solar wind. The layer below the corona is also heated and resupplies the corona with gas. (Photo: Hinode satellite, JAXA, NASA)

PART II

Dissolution and Horror

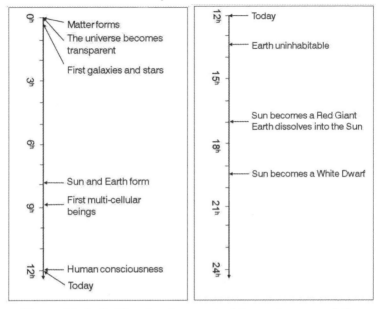

Figure 12: Left: The development of the universe and the solar system until the present in relation to the time from midnight to noon (13.8 billion years correspond to 12 hours). Right: The future development of the solar system using the same scale.

5

The Development Continues

The great astrophysical realization of the second half of the twentieth century was that the universe has a history. The cosmos was not created simply to remain in an eternally changeless state. Humankind would not have emerged in a universe without development, and we are a part of the cosmic story. But this story continues. After their formation, stars and planets develop further. Cosmic dynamics do not come to a halt.

5.1 The Young Sun

When the temperature in the center of a young star exceeds 20 million degrees Fahrenheit, hydrogen nuclei begin to fuse into helium. 570 million tons of hydrogen are fused in the Sun each second. The energy released in one second would satisfy humanity's present energy need for a million years. The fusion process in the center releases so much energy that the whole structure of the stellar interior and of its atmosphere changes. First, thermal radiation transports the energy outward. In the outer third of the Sun the temperature decreases so rapidly that the layers there become unstable: hot gas rises from the interior to the surface, cools, moves laterally, and sinks back inward. These convective gas currents thus convey heat to the surface.

The up and down of thermal motions in the outer part of the Sun has surprising effects. Since the gas has a temperature of thousands of degrees, the atoms move swiftly, and their collisions are so fierce that electrons are knocked from the atomic shells. In this way atoms become ions and, along with the ejected electrons, are electrically charged. Both react with the magnetic field. Charged particles are bound to the magnetic field like pearls on a string. Pearls can move along the string, but must stay attached. Pearls and string are a coupled system. If the string is moved, the pearls are carried along and vice versa. The magnetic string is elastic and acts like a rubber band, so that if the pearls move and stretch the string, the elastic force resists. Thus, the moving particles provide energy to the magnetic field and strengthen it enough that it finally dominates and contains the gas. In sunspots, gas motions have strengthened the magnetic field to the extent that all horizontal motion is blocked. Thus a sunspot receives less heat, stays cooler than the surrounding area, and therefore appears as a dark spot. It contains an unimaginable amount of magnetic energy.

When magnetic fields are carried by thermal movement to the solar surface, they extend into the atmosphere. There they gain the upper hand over the gas and dissipate. In this process energy is released, heating the dilute upper atmosphere, the corona, to more than a million degrees. Occasionally two oppositely polarized magnetic field lines meet and cancel each other. At that moment the magnetic field sheds its surplus energy explosively in the form of a flare. The flare energy can exceed the energy produced in a second at the Sun's center and is discharged within a volume equivalent to that of Earth. This energy release is not hidden in the interior, but radiates in all wavelengths, observable from gamma rays to radio waves. Gas particles in the corona are accelerated to more than half the speed of light. Some particles move back down to the surface and cause it to glow in X-rays. The rest escape into space.

It is not exactly clear how the corona is heated. Perhaps it is the many small flares occurring a million times per second all over the Sun that pour their energy into the corona. Other scenarios propose waves propagating from the bubbling surface. Whatever heats the corona, its gas is so hot and dense that the gravity of the Sun cannot hold it back. The outer corona flows as a solar wind past the planets into outer space. The mass loss by the solar wind is minute and, since the time of the Sun's formation, has not yet reached one-thousandth of the solar mass. Occasionally, entire portions of the corona are expelled because the magnetic field becomes unstable. These coronal mass ejections plow through the space between the planets, and their shock waves unleash storms in the solar system.

The outflow of the solar wind blows a cavity into the gas between the stars, which is called the heliosphere. The heliosphere shields the Earth from certain influences originating in outer space, such as high-energy particles called cosmic rays, interstellar gas, and interstellar shock waves. In the vicinity of Earth the solar wind still has a temperature of several hundred thousand degrees and is therefore ionized. Again the magnetic field acts like a string, but the particles from the Sun cannot attach to a string that is connected to Earth. The Earth's magnetic field deflects the solar wind so that it cannot reach our atmosphere. Even more important than this shielding from the solar wind, the terrestrial magnetic field holds at bay most of the hail of particles accelerated in solar flares. These ionized, high-energy particles, however, impact astronauts like radioactivity. They can be life threatening.

On the surface of the Earth we are well protected from the deadly intensity of space weather by the shield of the terrestrial magnetic field. Even the solar X-rays mentioned above cannot harm us. These would also represent a dangerous type of radiation for humans and would far exceed lethal doses if not stopped. This time it is molecules in the air that protect us, absorbing the harmful radiation high in the atmosphere. Only a small part of the ultraviolet radiation gets through and makes its presence occasionally noticeable in the form of a sunburn. Thus the surface of Earth appears as a unique place in the solar system where life can flourish—a well protected island amid the solar storms.

It wasn't always this way. The magnetic field, which is ultimately responsible for all solar activity, is driven by stellar rotation. Young stars rotate faster because a star inherits the angular momentum from the parental cloud core. Young stars with masses similar to the Sun are surrounded by coronas that glow a thousand times brighter and have flares ten thousand times more powerful than the Sun's. The young Earth was protected by a magnetic field and an atmosphere, but these shields were never perfect. The early Earth was exposed to a much greater bombardment by energized particles, and to a much more intense irradiation than is now the case. These must have had a significant impact on the early formation and evolution of life.

5.2 Design or No Design?

The jewel that is Earth, placed as it is in a treacherous universe, stirs biblical associations. Is there a plan behind this Garden of Eden in the desert of space? No! An actual blueprint, of the sort engineers and architects design before breaking ground is not conceivable in this case. A design is based on goals, decisions, and concepts. Good designs are not easily changed, once implemented. A design requires a clear route of development and somewhat stable conditions. This was not the case for Earth, since the Sun and its influence on Earth have changed drastically over time. Chance also played its role when asteroids and comets hit the planet.

Venus and Mars, our nearest planetary neighbors, are not suited for sustaining life. Venus, closer to the Sun, has a dense atmosphere rich in carbon dioxide. There the greenhouse effect drives the temperature above the boiling point of water and makes the atmosphere unfit for life. Mars, on the other hand, is too small to retain enough atmosphere and is therefore too cold today to maintain liquid water. If there were a design, then Venus and Mars would be examples of bad planning.

Design may be figuratively understood to mean a recurrent theme throughout the universe. Yet even as a metaphor, design implies an origin before the Big Bang when the design was established. The entire universe with all of its monstrous aberrations would thus be part of this plan as well. A metaphoric design does not easily fit the concept of a developing universe where many processes play off one another and where chance intervenes at important points. The situation becomes unsettled if the design is not recognized as a metaphor or as an interpretation. Then the notion of design could be misunderstood to claim the authority of scientific explanation. This is the case for the claim of Intelligent Design as it presents itself today.[1] It shortcuts the scientific method by giving rash and conjectural answers to scientific questions that are still open.

From a theological point of view, the notion of design also suggests a highly questionable metaphor, relegating God's interaction with the universe to the formulation of a single master plan in an inaccessible eon before time started. If something new were to appear in accord with such a design, it must have been anticipated as part of creation from the one design conceived beyond time and space. In contrast to this static view, the long-established theological notion of continuing creation (from Latin *creatio continua*) is better suited to describing a universe in the making. What this notion could mean will have to be clarified later. Continuing creation and design are mutually exclusive concepts. Creation indicates spontaneous formation through free will. It should be emphasized that neither creation nor design originate from scientific measurements or theories, but are metaphoric concepts describing a reality that is perceived in a participatory way.[2] How they may be useful in interpreting cosmic reality will be discussed later.

Metaphors used to interpret the universe are more than just rhetorical. They guide our perception and are, conscious or unconscious, important to our view of the world. The universe has been and will be interpreted metaphorically, but this kind of interpretation must be distinguished from scientific explanation.

5.3 The Decay of the Sun

The energy source that the Sun taps is hydrogen. Its fusion into helium provides the Sun with a stable equilibrium between gravity, which holds together the gaseous body, and pressure, which tends to expand the gas. The equilibrium is self-regulating: if the Sun cools, its diameter shrinks, density and temperature increase in the center, and the hydrogen atoms fuse faster. The additional heat increases pressure, expanding the Sun until equilibrium is reached again. This works as long as the hydrogen supply lasts. The stockpile of hydrogen is well known from the mass of the Sun. We also know the energy consumption from solar luminosity. Dividing the two values yields a finite life span. Stars that formed much earlier than the Sun indicate that hydrogen fusion accelerates during the lifetime of a star, and that not all hydrogen is ultimately consumed. The ashes of the fusion process, helium, initially do not react further and accumulate in the center of the Sun. To stabilize pressure, the core must become denser and produce more heat. The zone of fusion moves outward, expanding the body of the Sun, which then shines brighter. The Sun's luminosity has already increased 30 percent since its formation and will increase another 10 percent in the next billion years. In six billion years the Sun will be 2.2 times brighter, and Mars will be illuminated as much as Earth is today.

In seven billion years, solar evolution will dramatically

accelerate. The Sun will expand to a Red Giant twice within a 150-million-year period. The surface will cool and become red. Older solar-like stars in the galaxy show that the Sun's diameter will increase two hundred times, and that the luminosity will exceed the present solar value by a factor of a thousand. The Sun will expand out to the present radius of the Earth's orbit around the Sun. There is no hope for Mercury and Venus; the Sun will swallow them. What will happen to Earth? When the Sun expands, the solar wind will become stronger. It will carry away a quarter of the Sun's mass into space. Solar gravity will diminish, increasing the orbital radii of the planets. In its future orbit, about where Mars circles today, Earth will attract solar gas and lose orbital energy. According to the latest calculations, the Earth will be unable to escape the expanding Sun.[3] It will gradually sink toward the Sun and finally evaporate. In 7.5 billion years its atoms, including all those present in human bodies, will intermingle with those of the Sun.

Earth will become uninhabitable long before that. Already in 1.2 billion years there will be places on Earth where the temperature will exceed 212 degrees, and water will evaporate in large quantities. Water vapor is an efficient greenhouse gas if abundant in the atmosphere. Its effect will exceed that produced by humanity's discharging of carbon dioxide. As soon as there is enough water vapor in the air, the temperature will increase even more, eventually evaporating all remaining water. In the upper atmosphere sunlight will split water molecules apart, allowing single atoms to escape into space where they will be lost forever. After this last great greenhouse catastrophe the soil and the oceans will be completely dry. Earth will become hot and forbidding like present-day Venus.

When the temperature in the Sun's interior exceeds 200 million degrees, the first Red Giant phase will end abruptly. The pressure will then be great enough for helium to fuse into beryllium, carbon, and oxygen. This new energy source brings the Sun into a new equilibrium. The solar radius will still be ten times larger than it is today, but for the next 100 million years peace will be restored to the solar system.

Helium burning does not release as much energy and therefore does not last as long as the hydrogen phase of the Sun. Once again the structure of the interior changes and the Sun becomes a Red Giant for the second time. During this time, the Sun pulsates and eventually reaches beyond Earth's present-day orbit. Gravity at the surface becomes so low that the outermost layers of the Sun slip away in the form of an extremely strong solar wind. Shells of gas drift into interstellar space, forming what is known as a planetary nebula. Its brilliant colors will be visible throughout the galaxy for more than 100,000 years. If some alien civilization actually sees it, it will know what happened to the Sun and its planetary system. The development of the Sun may knock the planetary orbits out of equilibrium, which could bring danger to small survivors like Mars. If Jupiter or Saturn deviate from their present circular orbits to more stretched-out ellipses, Mars could either get lost in the Sun or be propelled into space. The fate of Mars is not yet established.

Computational models assign the Sun a lifetime of 12.36 billion years.[4] There are still 7.79 billion years until the nuclear energy is exhausted. Then the Sun will shrink to a White Dwarf and slowly cool. Its diameter will be little more than the Earth's and its mass reduced by half. Initially, the solar White Dwarf will glow white-blue, later becoming red and diminishing in luminosity. The temperature in what is left of the solar system will equal the coldness of space.

5.4 Becoming and Decaying

What is born will perish. Decay is inevitable. Stellar evolution does not end with formation. Since energy is released continuously, it will one day be exhausted. Formation seamlessly turns into dissolution. Decay is the counterpart to

formation. Creativity in the universe also implies the passing away of all things. The optimism inherent in the models
of star and planet formation, as well as in biological evolution, stands in marked contrast to the sinister expectations of
the future. We have to accept that we are living in a universe
where all material things will sooner or later decay.

If becoming and decaying belong together, the notion of
creation must also include its opposite. This principle was
forgotten in the ideal world of physico-theology. Yet the last
chapters of the Bible portray the end of the world in dark
scenes. Humans in earlier times felt closer to the abyss of
collective termination. The connection between creation and
death was recognized as a matter of course. Whether emergence or decay is perceived to be more prominent depends
largely on a person's phase in life and temperament, or on
the current societal mood.

The Sun will burn out only in the inconceivable time span
of billions of years. Nevertheless, the shadow of transience
must be seen as blended with the sense of wonder about
the finely woven web of processes forming stars and planets. The alleged permanence of the cosmos is gone and with
it all of the world's stability. There are no eternal objects
in the universe. When I, as a student, first became aware
of the finite lifetime of the Sun, I was more relieved than
afraid. It became clear to me that no human achievement,
however great, would have infinite duration. All that we
achieve in our lives will pass away, probably even scientific
knowledge. Our work will not achieve immortality, which
therefore cannot be the goal of our research.

To pass away does not mean that objects disappear into
nothingness. For as we have seen, the end of massive stars
sets the stage for planet formation. There is a parallel to this
in biology: evolution can proceed through the death of individuals, allowing a species to adapt to a changing environment and to survive. The universe can only develop because
material things are not eternal.

6

Living in the Midst of Evolution

The Earth has a deep history of development and an expansive future. But let's return to the present. Even today the Earth is developing. Life on Earth is a part of this history and actively participates in the planet's changes. Humankind's destiny is played out in the midst of cosmic evolution. Our habitat changes continuously, and is therefore also in danger. We are on a path that is uncertain in many ways. Along this path lie legions of the deceased that perished as a result of changes on Earth. They carried the burden of evolution and so ensured the existence of animals and humans on the planet today.

Changes started already when the Sun formed. Stars, as described on page 18, are a group phenomenon. They do not form alone but in clusters. The Sun, too, came into existence within a molecular cloud. Such clouds are kept together by their own gravity. When a star cluster emerges in the cloud, the stellar radiation and winds dissolve the cloud. Since only a small fraction of the gas ends up in stars, much of the initial cloud mass is expelled. The reduced gravity of the newly formed star cluster may not be sufficient to keep it together. In such cases the cluster dissolves and the stars disperse over the galaxy.

The Sun's history followed this same trajectory. Its molecular cloud disappeared long ago. Thousands of its stellar siblings are scattered all over the Milky Way and, at such great distances, may be identified only tentatively. The molecular cloud must have been huge, such that also massive stars formed. They passed rapidly through their lifetime when our molecular cloud was still intact. Once a supernova exploded close by in our neighborhood and left its traces in the form of gold, lead, and uranium atoms that enriched the cloud core from which Sun and Earth formed. Uranium isotope ratios permit precise radiometric dating of the time of the explosion. It approximates the age of the solar system. The abundance of uranium indicates that the exploded star was only five light-years distant from the Sun.[1] The resulting shock wave reached the site of the future solar system within a hundred years and may have triggered the collapse of our cloud core and initiated the formation of the solar system.

6.1 The Young Earth

Earth-like planets accrete from dust that formed in stellar winds and supernovae ejecta. Dust grains start as loose structures only a few thousandths of a millimeter in diameter. When the cloud core cools, they develop a mantle of water ice. It is far from clear how planets form out of such grains. One of the scenarios envisions the process in this way: dust grains move in the gas of an accretion disk with different velocities. The larger a grain, the faster it must move relative to the gas and the more frequently collisions with other dust grains occur. The sticky ice mantel acts like a glue, thus preventing the colliding grains from disintegrating in the process and allowing them to continue growing. Planetesimals of one-yard size begin to form. These are the predecessors of planets. Big planetesimals move through a myriad of smaller bodies that stick to them, sweeping clean the space within their reach. Large bodies grow even faster by gravitational attraction, so that their growth becomes unstable.

The growth of a planetesimal continues in this way until its gravity becomes strong enough to influence the orbit of neighboring bodies. Then a stormy phase of evolution begins. When planetesimals interact, their orbits deviate from the initial circular and become elliptical. Orbits cross one another and planetesimals collide with speeds up to 40 miles per second. In such collisions chance reigns; the two planetesimals may break into thousands of fragments or become hot enough to melt together. Again evolution proceeds such that large bodies consume the smaller ones.

The first 100 to 200 million years of our planetary system must have been very eventful. About 40 million years after its formation, a dwarf planet hit the Earth. The object—called Theia, after the mother of Selene, the Greek goddess of the Moon—was as large as Mars. Instead of striking Earth's center, Theia only grazed the outermost layer and melted completely. The energy of the encounter was unimaginably large. Vaporized rock from Earth and Theia mixed and was ejected into space, forming a ring around Earth. Within less than a hundred years the material coagulated and formed the Moon. Support for this theory comes from rock samples brought back by the Apollo astronauts from the Moon between 1969 and 1972. They indicate a surprising similarity between Earth and Moon. The collision made the Earth's spin increase significantly. After the impact, the length of a day was only five hours. Tidal interactions with the Moon over billions of years slowed down the rotation rate of the Earth to today's 24 hours. If the clash had been head-on, there would be no Moon today. Earth would have heated up such that carbon dioxide was enriched in the atmosphere, enhancing the greenhouse effect.[2] Temperature on Earth would be substantially higher even today, creating an atmosphere as hostile as that on Venus. When I look at the Moon, I am sometimes reminded of how the unique environment that supports humankind was shaped by many strokes of good luck.

Collisions are important and completely normal events in the course of planetary formation. They ensure that not too many planets survive, thereby stabilizing the overall growth of planetary systems. Impacts during the Earth's early history added material that enabled the planet to grow considerably over millions of years. At the same time these impacts heated Earth to the melting point. Heavy elements such as iron and nickel sedimented toward the center of the Earth, while lighter elements, such as silicon and aluminum and their oxides, formed the crust. When the impact rate decreased after 200 million years, Earth cooled so that liquid water probably existed, making life possible in principle at this early time.

Collisions with smaller bodies again became frequent another 600 million years after the formation of Earth. It was the time of the "Late Heavy Bombardment," when the orbit of Saturn around the Sun shrank, making the planet's orbital period twice that of Jupiter, therefore placing both planets into resonance. The mutual attraction compounded so that Saturn gained a bit of energy and Jupiter lost a bit with every encounter. Their orbits became elliptical. Many small planetesimals were pushed into even more elliptical orbits that also brought them into the inner solar system. For 300 million years, mile-sized objects hit Earth about every twenty years, leaving craters of more than ten miles in diameter. A large part of the impressive cratered landscape of the Moon originates from this time. The Earth's crust was still thin, so that large objects penetrated the interior and left no traces visible today. Shortly after, life appeared on Earth.

6.2 Earth Is Still Growing

Even today, several thousand tons of remnants of our former accretion disk fall to Earth per year. Most are dust grains weighing less than a milligram. Meteors in the sky are their luminous traces. Most of the material originates from collisions in the asteroid belt. One-third of the meteors that reach the surface today, meteorites of the L-chondrite type, are derived from a collision 470 million years ago when a 100-mile-sized asteroid was blown apart.[3] Debris, but also other asteroids, comets, and tiny bodies of dust grains, move on elliptical orbits across the nearly circular orbits of the planets. The orbits intersect, and collisions and impacts occasionally occur. The impact rate of large bodies may be calculated from crater counts on the Moon, Earth, and Mars. Except for occasional large events, the rate has been nearly constant in the inner solar system for three billion years. The probability of an impact may therefore be determined accurately. On average, Earth is hit about once per millennium by an object as large as 50 yards in diameter. If the impact occurs on land, an area fifty miles in radius is devastated. If the impact is in the sea, a tsunami may ravage the nearby coasts.

The small bodies outnumber large ones. Thus, it is much less likely that a larger object will fall. For instance, a kilometer-sized meteorite striking the Earth, like the one that hit the estuary of the Chesapeake Bay, is expected once every three million years. The energy of such an impact equals the blast of 100 gigatons of TNT, about seven million times that of a Hiroshima bomb. Dust enters the atmosphere and cools the Earth's climate for years. More than 900 asteroids of this size and larger are known that cross the Earth's orbit. Some of them will impact the Earth or the Moon. The chance that an object of this size will strike in the next thousand years is one in three thousand. The asteroid Eros, for instance, is a well-known candidate. It has a diameter of ten miles and a 5 percent probability of impacting Earth in the coming 100 million years.[4] The catastrophe would be larger than the one 65 million years ago, during which time the dinosaurs became extinct.

Large impacts shoot a fountain of hot steam, liquid rock, and small dust grains into the atmosphere. A hail of boulders hits a large part of the Earth's surface and distributes the energy of the impact over all continents. The air is heated by several hundred degrees, igniting vegetation. Fire spreads over the globe; smoke, ashes, and dust change the climate and the soil. Meteorites larger than ten miles punch through the Earth's crust. Liquid rock pours out from the hole and floods a region as large as a continent. After the largest extinction event, possibly caused by volcanic activity 250 million years ago, it took a surprising five million years until trees grew again in Europe. The recovery time of the plants after the event 65 million years ago was still tens of thousands of years. The dinosaurs did not outlive it.

Over the last decade, the sky has been carefully searched for large asteroids. The most dangerous asteroid found so far is 2004MN4. Its name includes the year of its discovery. At first, astronomers were greatly concerned, as observations over several months yielded orbital predictions that 2004MN4 would approach Earth dangerously close in 2029. The probability of a collision was estimated at 2 percent. The asteroid was renamed Apophis, after the evil god of darkness and chaos in Egyptian mythology. The impact of this 300-yard rock would be disastrous for a continent and have worldwide ramifications.

Whether a body will actually hit the Earth remains uncertain for a long time because its position and motion cannot be determined with total accuracy. Be it the air turbulence that blurs the image or the size of the telescope that restricts the resolution, there is always a limit to how well something may be observed. If position and velocity are not exactly known, the orbit cannot be exactly predicted. The inaccuracy increases with time. Instead of a line, the predicted trajectory takes the shape of a hose, limiting its probable course. The thicker the hose, the more uncertain is the predicted orbit. The hose widens into a cone in the direction of forward motion. More precise measurements shrink the cone. That was also the case with Apophis. Within a month of its detection, the largest radio telescope in the world in Arecibo (Puerto Rico) pointed its 1,000-foot-diameter radar dish to the asteroid and determined the orbit with higher accuracy. The result was that Apophis is not on a collision course with Earth, but will pass it at a distance of 20,000 miles. It will be an object visible above Europe, Africa, and Western Asia to the naked eye for 40 minutes at one hour past sunset on Friday, April 13, 2029. It will cruise between Earth and the geostationary satellites with a velocity of four miles per second. What kind of experience will it be to witness it?

Apophis will remain a near-Earth asteroid whose orbit crosses that of Earth. The object will approach Earth again in 2036. Accurate predictions of its orbit then are impossible since it will be so strongly affected by its 2029 encounter with the Earth. The danger is small but real.

The example of Apophis demonstrates that the orbit of a planetary body cannot be calculated far into the future. Sun and planets shape the orbit of an asteroid. Yet, they attract one another as well, and together form a nonlinear system. An error, inevitably arising from inaccurate initial values, usually grows exponentially: slow in the beginning, then faster and faster. This means that after a certain time—the Lyapunov time—the uncertainty increases such that an even better initial value does not yield a significant improvement of the predicted orbit. To know the position of Apophis in seven years twice as accurately, its present position would have to be known a hundred times more precisely. This behavior is called "chaotic" because the far future of the system is not predictable. The useful prediction time for many asteroids is only a few hundred years. In a close encounter with another body, a difference of only a few miles significantly influences an asteroid's future orbit and drastically reduces the predictable time. Chaos in the solar system is a remarkable property with which we have to live. It means, among other things, that it is not possible to compute on which date the next large object will impact Earth. We have to be content with probabilities in predictions involving the distant future.

It is beyond doubt that asteroids and comets are currently heading toward Earth. Already today, our high-technology civilization could deflect an asteroid to an orbit away from Earth with a space mission. Several methods have been considered. Yet there are so many asteroids smaller than a hundred yards in diameter that it will never be possible to identify all of them. An impact may occur without a long lead time. The same holds for comets arriving from the outer solar system. They become visible only when they cross the orbit of Saturn, at which time a few months remain to act, perhaps too late for an emergency mission.

6.3 Evolution of the Atmosphere

The orbits of nearby stars are also chaotic. For this reason it is not possible to make long-range predictions about which star will approach the Sun or which might even explode as a supernova in our neighborhood, an event that would change the terrestrial environment for years through irradiation. All supernovae that exploded in our galaxy during the past millennium have left traces in the ice of Antarctica. High-energy X-rays and ultraviolet radiation emitted by a supernova hit the upper atmosphere of Earth and induce chemical reactions that produce nitrates. They are brought to the ground by rain and snow, and are deposited in the ice.

To cause a catastrophe similar to a large meteorite impact, the supernova explosion would have to be closer than 25 light-years. This happens on the average less than once per billion years. There is a sediment layer in the sea with enhanced abundance of iron 60, a radioactive isotope that traces a nearby supernova explosion. It may be dated at two million years ago and may possibly be associated with a supernova remnant in the Scorpio-Centaur star cluster, about 120 light-years away.[5] There is evidence for a small mass extinction event at this time.

Earth is not only endangered from the outside. Its own development continuously changes the environmental conditions of life. Such an incident occurred from 800 to 600 million years ago. At that time protozoa comparable to today's bacteria were the dominant form of life, although multicellular life forms also existed, having originated some 100 million years earlier. There was no life on land. All organisms lived in water and did not exchange molecules with the air. Yet the atmosphere of the Earth was not completely static. It changed slowly by interaction with water and rocks. The atmosphere at this time contained little oxygen, but the carbon dioxide formed carbonates that were deposited on the floors of the oceans. When the concentration of carbon dioxide fell to less than half of today's level, the greenhouse effect diminished. In addition, the Sun was 6 percent less luminous than today. It became cold on Earth, and water in the atmosphere increasingly precipitated to the surface. When large areas were covered with snow and huge glaciers advanced toward the equator, there was soon hardly any water vapor left in the atmosphere, making the sky cloudless. More and more solar energy radiated from the snow-covered surfaces into space. The Earth cooled further until it finally became a snowball.[6] Temperatures on the surface dropped to an average of -17 degrees Fahrenheit. Lakes and even oceans were completely ice covered.

Organisms could survive only in the few ice-free areas of the seas near the equator. Thanks to these refuges, life did not become extinct on Earth. Life under these extreme conditions was exposed to heightened evolutionary pressure. This largest ice age ever was probably the reason why multicellular creatures developed and became much more widespread when the catastrophe was finally over. It was the time of the so-called Cambrian Explosion, when many new species of multicellular beings originated and from which time the genetic blueprint of many phyla date. At this time many animals, for example mussels, developed a hard outer skeleton of calcium carbonate for protection. Gigantic volcanoes, possibly related to the breakup of the supercontinent Rodinia, the only large land mass on Earth at the time, outgased new carbon dioxide into the atmosphere. The greenhouse effect increased and finally ended the climate disaster.

6.4 Life Is a Risk

When an electric field builds up in a thundercloud, the voltage is far below the threshold needed for a discharge between cloud and ground. Usually air does not conduct an electric current. The phenomenon of lightning obtained an unexpected explanation a few years ago, when geophysicists of the Tieng-Shang Observatory in Kazakhstan were surprised to observe lightning strikes in thunderstorms often exactly at the time when their particle detector registered a cosmic ray of extremely high energy. Cosmic ray particles are mostly protons, the atomic nuclei of hydrogen. They are accelerated to high velocities at the shock fronts of supernova ejecta and spread over the whole galaxy into the remotest corners of molecular clouds and to the surface of Earth.

The scientists in Kazakhstan began to measure both simultaneously and found that for every strike of lightning there were traces of a nearby cosmic particle. Apparently, the subatomic particles trigger the discharge in thunderstorms.[7] When the particles hit our atmosphere after a thousand-year journey through the galaxy, they collide with molecules in the air, knocking electrically charged electrons from them. This happens in a large number of collisions, and soon an avalanche of electrons builds up along the path of the cosmic proton. If a particle hits a thundercloud, the electrons react to the electric field and are accelerated. They also collide with further air molecules and knock out yet more electrons. Finally, an electrically conductive channel forms in which current can flow, providing a pathway for the lightning strike. Thus a single cosmic particle from a far away supernova can, millennia later, become the trigger for a fatality on Earth. We are surrounded by cosmic history wherever we look. The universe has a direct impact on our lives. From this perspective, there is no longer a distinction between above and below. We are in the midst of it all.

In addition to external influences there are terrestrial catastrophes. Large landslides into the sea and earthquakes threaten coastal inhabitants with tsunamis. In this century we may not be spared from environmental and biological disasters, such as erupting volcanoes, global epidemics, and world wars. But their occurrence is unpredictable. Although some laws of science are fixed, the evolution of the Earth remains open ended.

Earth is part of the universe and completely embedded in its development. Thus it should come as no surprise that the universe also plays a role in biological evolution. When the dinosaurs died out in the course of a meteorite impact, mammals exploited the ecological niche and took the dinosaurs' place. Two million years ago the *Hominidae* family had a boost in evolution and the genus *Homo* appeared. These momentous changes may have been driven by the supernova that exploded nearby, or a one-mile-diameter asteroid that hit the South Polar Sea.[8] By virtue of intelligence and mobility, *Homo* survived both crises with fewer adverse effects than many other creatures. Had such cosmic catastrophes not occurred in the past, humanity as we know it would not exist today. Nonetheless, that which seems to constitute a minute threat to an individual human today is almost certain to endanger humanity in the future. Life has endured on Earth thus far, but not without massive losses.

Life is a risk, but each risk is also a chance. We experience the same in personal life. Here we can discern a parallel between cosmic evolution and a characteristic of our own experience. Without being asked, human beings are born and find themselves placed in a given environment. We had no choice about the place and time of our birth. We were cast into a history that developed over billions of years and that will continue long after our death. Our habitat has certain distinctive assets and liabilities. The key point, though, is that it remains in constant flux. The changes we must endure alarm and threaten us, but they also yield new possibilities.

What holds true for the individual also holds true for all humankind across a broader span of time. The rapid development of the genus *Homo* two million years ago did not take place under stable conditions, as had been assumed until recently. The asteroid impact and the supernova significantly altered Earth's climate twice within less than a hundred thousand years. The environment changed rapidly, thus imposing drastic new demands on organisms. The universe is not a clock that ticks quietly into the future, but an adventure.

The evolution of the universe, the origin of life, and the existence of each human being share the commonality of an open future. This openness does not mean that just anything can happen. There is a framework, dictated by laws, within which much is possible so that new things can and do emerge. To an appreciable degree, also the course of our lives remains open to the future. We plan under the assumption that no unexpected changes will occur. However, that cannot be the case forever. We cannot reliably predict our lives. The reason is not just deficient knowledge; rather it seems that what will be tomorrow simply has not yet been determined today. This recognition within our common experience resonates with the "reality" of quantum mechanics, which is not discernible until proper measurements are made.

The uncertainty of development raises the question of how and why one of many possibilities ends up materializing. Science is often expected to provide the answer to this question concerning the cosmos. Can it be assumed to be the key to unlocking the essentials of reality also in life? This function was previously, and still is by some today, exclusively attributed to religion. In the following chapter, then, we need to pursue further the question of what reality is and how it might be perceived.

Figure 13: Massive stars formed in the center of the molecular cloud RCW 49 (bright spots in the middle of the picture). They blew a cavity into the cloud. More than 2,200 stars have formed in the entire cloud. This image in infrared light shows how the cloud is heated and decays. RCW 49 is at 14,000 light-years from us and has a diameter of 350 light-years. (Photo: E. Churchwell, NASA, Spitzer, JPL, Caltech)

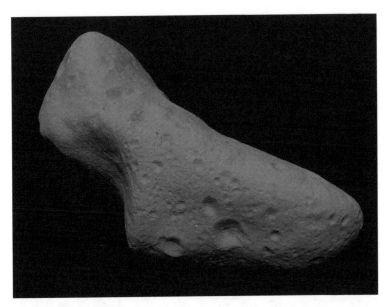

Figure 14: The asteroid Eros was approached and investigated by the NEAR mission in February 2000. Eros is 22 miles long, weighs a trillion tons, and rotates once in five hours. The impact craters suggest an old age. Today, the asteroid is not a near-Earth object, but a close encounter with Mars could bring it into a collision course with Earth. (Photo: NEAR Project, Goddard SVS, and NASA)

7

Reality in the Cosmos and in Life

When I was a student, I always looked forward to having dinner with my professors at the end of each semester. There was time to talk informally about everything and to ask questions one would never ask in a classroom setting. The dinner was often combined with a little excursion. Once the trip took us to the castle of Laufen and nearby Rhine falls in northern Switzerland. Before the meal we walked down to the river. A thin veil of droplets rose from the falling water. On a viewing platform at the very edge of the thundering waters I asked one of my professors, from whom I had just taken a course on statistical mechanics, whether it would be possible to calculate the trajectory of each droplet so that we could accurately predict how the falls would look after dinner. After pondering the question for some time, he answered that, yes, it should be possible. The water's motion, he argued, is after all the result of physical forces, and thus determinable and computable in principle. Somewhat surprised and incredulous, or perhaps because I thought it more interesting to offer a different opinion, I asked the provocative question of what "in principle" means in this case. What might our inquiry have to do with reality if we know it only in principle, but not in fact or experience and not in our personal reality? The question launched a discussion that lasted throughout the meal.

Science aims to explore reality, but not all of science concerns itself with what is demonstrably real. That which is observed, measured, and confirmed qualifies as fact. If the observer has not made a gross error, the resulting data are immediate reality. That is not the case, however, for whatever theories, hypotheses, and conjectures are developed from the data. Such constructs are preliminary, may be contradicted by new measurements, or may need to be adjusted. Yet it is precisely these explanatory constructs, rather than measurements, that ordinarily influence our worldview, shaping our perceptions and expectations. In turn, we perceive most clearly that which we can most readily integrate into our worldview, and happily repress the rest.

Sooner or later in our lives another reality catches up with us. In my daily life I perceive much in a very different way than I do through science. Human beings experience as real those things that have a lasting effect on them. Emotional experiences like art or love have existential importance. Happy events, illness, and death are characterized by a depth that transcends any simple causal explanation. That which we experience as real is more expansive than what science can perceive, and often on a different plane. To comprehend the furthest horizon of human reality, we must extend our journey into the universe full circle to inquire about ourselves, and our discrete experiences of what we take to be real. They, too, are part of the universe.

Accordingly, this chapter deals with how the several levels and planes of reality confront us, in diverse ways, as mediated through multiple branches of science as well as through human experience.

7.1 Levels of Reality

I am amazed by how much physics can explain with just a few fundamental formulas. With nearly identical equations, physicists describe the motion of droplets in a waterfall as well as the dust grains and gas molecules in an interstellar cloud. In both cases there are at least two levels of accuracy with which reality is measured. At the lowest level, researchers can focus attention on single particles such as droplets, dust grains, or molecules. Their orbits and collisions with other particles follow the fundamental equations of physics. In view of the large number of particles (some 10^{64} in a molecular cloud), this information is much too detailed for all purposes. Therefore gases are usually described in physics on a higher level, where only the mean number of particles, mean particle velocity, and mean particle energy are of interest. Equations may be found for these mean values that allow the development of the gas to be described. Physics at this level does not deal with individual particles but rather with collective characteristics such as gas density, gas speed, and temperature at a certain location. Although these concepts do not make sense for single particles, they describe an authentic reality. This example makes clear how, even in physics, a further reality can be grasped by declining to focus on the smallest components of what is observable.

The tangle at the lower level is summarized by a few material constants, such as heat capacity or electric conductivity. The constants are different for different gases. They may be measured or calculated. Such calculations assume that the orbits of individual particles, their positions and velocities, are random around certain mean values. Thus the interactions and collisions are also random. This methodology describes the field of statistical mechanics. It explains the observed mean values using the theory of random particle distributions.

Randomness is not always the norm. The human brain is a counterexample. Neurons cannot be described as randomly interacting particles. Rather, there is a purposeful interaction of certain individual cells among an immense number of neurons. An average value of all cells tells us nothing of this interaction. In the specialized cases where interactions occur, the processes are not random because it is the individual that counts. Here statistical physics reaches its limits. Nevertheless, biologists find regularities in these neurophysiological processes. These patterns do not contradict the laws of physics that govern each cell. What's more, they also operate according to laws at a higher level. This principle explains why biologists do not need to study quantum field theory or the standard model of particle physics to conduct biological investigations. They employ other methods and concepts than physicists do to study the phenomena at work in living organisms.

These examples illustrate how reality is variously perceived and explored in different fields of science. A reasonable choice of concepts, examined at a higher level of complexity, is the key to successful research in these fields. Avoiding an accumulation of excess detail on the lower level is to the best advantage at the higher level. That which applies to statistical mechanics and biology applies as well to other research disciplines, including psychology and sociology.

There is yet another dimension of reality, different from the levels examined by various branches of science. I want to suggest that the knowledge gained through scientific exploration is only a part of all reality. We are dealing here not just with a linear extension or multiplication of the levels described above, but with a qualitative step toward another perception of reality altogether.

7.2 Religious Perceptions

After I had presented an interdisciplinary lecture on science and religion, a graduate student in astrophysics came up to me and said that he couldn't think of anything that doesn't follow Schrödinger's equation. How, he wondered, could a physicist like me imagine something beyond this fundamental equation of quantum mechanics, which describes in principle every temporal change that is possible? This polite challenge exposes a widely accepted worldview in which science—physics above all—is presumed to be the basis of reality, from which all that is real derives its footing. But has modern science established definitively that the sole ground of reality is that which physics can measure and explain? I do not believe so, particularly if we reflect on what a fully comprehensive concept of reality must encompass.

The methodology of physics starts with the practice of measuring or observing a certain phenomenon. The measurement must be made in a way that can be repeated by anyone at any time. Such a result is said to be objective. Results are bound together with mathematical formulas to create theories. Theories are models of reality. With the benefit of new measurements, their congruence with reality can be checked and improved. Through repeated cycles of measurement and theoretical explanation, physics enlarges and deepens its range of reality. All other sciences practice a similar methodology.

Objective measurement demands more than intellectual honesty. Effective methods of scientific measurement also require discretion about which aspect of reality is properly subject to investigation. For scientific study, only phenomena that can be measured objectively may qualify. It is this selection that makes science successful. How could a machine work if its basic principle of operation were unpredictable? How else could generally valid principles of star formation be found amid the confusion of molecular clouds? Reducing the field of investigation to objectively measurable perceptions limits science quite critically at its outset.

In effect, the graduate student's question was whether perceptions exist that must be taken seriously but do not fit the criteria of physics and so cannot be explained, even in principle, by Schrödinger's equation. The limits of any branch of science are defined at the beginning by its methodology, assumptions, and procedures. Measurements are made and observations are selected according to these rules. Given these constraints, it is not possible to judge scientifically the existence or character of any reality beyond a given field of science. Only human perception and experience, not scientific theory and method, can access the full range of reality open to humanity.

Perceptions are externally related influences that have become part of our consciousness. They include but are not restricted to scientific measurements and observations. Different kinds of perceptions together constitute our window onto reality. Forms of perceptual reality beyond the limits of a given branch of science must not, however, be denied on principle. Refusing on narrowly methodological grounds to consider the full scope of reality threatens, ironically enough, to subvert the scientific ideals of the Age of Enlightenment. After all, a major virtue of the modern scientific method is its unbiased perception of the world.

One example suffices to demonstrate the incompleteness of science. The experience of Blaise Pascal (1623–1662) on the night of November 23, 1654, 10:30 PM to 12:30 AM, is particularly noteworthy. It is a mystical perception, a type of experience that has become unfamiliar in our culture. The renowned mathematician and physicist dated it exactly, wrote it down in fragmentary words, and sewed it as a "memorial" into his clothes. There it was found after his death. The mysterious words read:[1]

> *Fire.*
> *"God of Abraham, God of Isaac, and God of Jacob"*
> *not of the philosophers and the learned.*
> *Certitude. Certitude. Feeling. Joy. Peace.*
> *God of Jesus Christ. . . .*
> *Forgetfullness of the world, and of everything, except God . . .*[2]

The stammered words apparently describe a religious perception (fire), and a first attempt at its interpretation

(God of Abraham . . .). We cannot know exactly what Pascal perceived, as the experience is tied to his person (certitude, joy . . .). We can only compare it to the perceptions of other people or to our own experiences. This is what Pascal did. He linked his perceptions to the heritage of the Old and New Testament.

A comment here regarding the notion of a "personal God" may be appropriate. This metaphoric concept of transcendence, much criticized among physicists,[3] is hinted at in Pascal's reference to the biblical God. Can the divine, which is imagined to encompass and even exceed the unimaginable dimension of the universe, have the traits of a person? Personhood in this case is clearly to be interpreted as a metaphor, one that apparently accounts for Pascal's shock when he encounters the person-like divine that he also describes in figurative terms as fire.

The mystical experience had a decisive influence on Pascal's life. It shaped his thought and provided him with a fundamental assurance that no one could question. What does this perception have to do with reality? Was it an illusion? Illusions do have an effect, but their influence cannot ordinarily be expected to endure. Pascal's perception did have a lasting effect. It is therefore appropriate to speak of it as engaged with reality, though this reality cannot be verified by others.

Since the eighteenth century, claims have been circulating that in the not-so-very-distant future science will explain everything. Whether such will ever occur remains to be seen. The "everything" in question, however, can only refer to the limited range of scientific perceptions. We do not know how far this range extends. It is indeed remarkable how successful the scientific method has proved to be, and how far it has taken us in astronomy, physics, chemistry, biology, and neurology. Such achievement is impressive, despite the errors and incomplete explanations that scientific inquiry produces along the way. Reli-

gious perceptions involve neural activity, but are not susceptible to empirical tests and are clearly not the basis for
scientific inquiry. As such, they should not be expected to
figure in the course of scientific investigation of any kind.
Pascal declined to discuss them in the context of his physics research.

7.3 Participatory Perceptions

Whitman's experience, described in his poem about stars
(see p. 52), is another example of a perception that sur-
passes the reach of scientific measurement. On a clear
night the retina of the eye registers light in the form of
photons that are emitted by stars. People react differently
to these photons, whose wavelength and intensity can be
measured with a spectrometer and secured in numbers.
Yet we may also allow ourselves to be moved as Whitman
by the presence of these celestial bodies. We can feel a certain connection with the stars and "hear" their silence. We
know that the light we are witnessing has been traveling
for many years, with a travel time we may contrast with
our own lifetime. We know, too, that the temperature of
stellar atmospheres radiating this light is thousands of
degrees and that the light we witness must have traversed
an unimaginably vast space of life-forbidding emptiness.
Reflecting on these things, we stand amazed that our environment, with its moderate temperatures, can exist in this
universe at all, or that we ourselves exist. And as the horizon widens, we perceive ourselves as a part of something
still larger. As though a veil had fallen from our eyes, we
realize how insignificantly small we are, how short our lives
are, and yet how beautiful is this planet Earth we inhabit.
We imagine ourselves wandering through the remotest
regions of the universe, reflecting on far-distant sites we
no longer consider as wholly unknown. Such intuitive
search for one's place in the cosmos is beyond the range of

objective science yet represents a form of experience that helps to orient us and that holds existential meaning.

Another kind of participatory perception is illustrated by our experience with art. We go through an art gallery and stop at a certain picture. It appeals to us. We may proceed to the next picture, or remain and involve ourselves in a kind of dialog. The light that reaches an observer from the picture originates either from a lamp or the sun, but is reflected only in the colors of the paint. The light reaches the retina, and signals are transmitted to the brain. So much for science. In our consciousness we feel an immediate presence of the image on the wall. An art experience is not objective. Though it requires an outward-directed focus of the viewer's gaze, this phenomenon can occur only if the beholder participates. He must take his time and concentrate on the picture. Whether or not the image appeals to the viewer depends on something more than individualized subjectivity, since there are pictures to which more people react and which therefore command higher prices on the art market. The experience amounts to a resonance phenomenon in which the outwardly observed piece of art provokes an inward response in the beholder's consciousness.

In physics, a passive object is said to resonate if it is gripped by a wave and begins to vibrate with it. The violin bow, for example, provides the energy that causes a string to vibrate. The string alone would hardly be audible. Its oscillations are transmitted by the bridge to the body of the violin, which is the resonator. The whole violin now starts to oscillate. It captures part of the energy and transmits the oscillation with its much larger surface to the air so that a sound wave is emitted loud enough for the human ear to perceive. Similarly, an experience in visual arts needs a picture (corresponding to the vibrating string) as well as a person (the resonator).

A person participates differently in this second kind of perception than in one involving scientific measurement

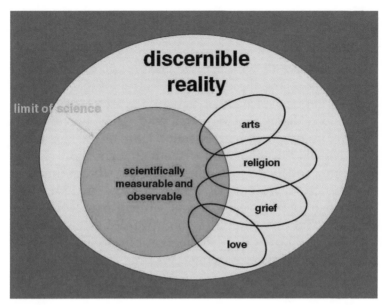

Figure 15: Schematic of human perception of reality. Red: subset that science can objectively measure or observe. Light blue: what we can discern by participating. Ellipses indicate particular perceptions, for instance of art, which involve participation, but include measurable processes in the brain or in the blood. The same applies to religious perceptions, the feelings of grief or love. The essence of participatory perceptions is not part of science.

and observation. The human being, with his consciousness and emotions, is the measuring device, rather than
the telescope and the camera. Thus, I refer to this personally involved response as *participatory perception*. With this
personal involvement, a subjective element is introduced
into the perception of reality. But is this kind of perception
merely fanciful, subjective, or illusory, the product of way-
ward brain activity? Such a dismissal seems unwarranted.
Granted, participatory perception may scarcely be considered apart from the subject viewer, and the person involved
may not at any moment be particularly conscious of his or
her subjective investment in the process. But there is also an
interactive dynamic at play here, in which the human per-
ceiver—rather than any measuring device—serves as the
actual instrument of sense. True, one can decline to engage
in participatory perception, the equivalent of walking by a
painting in the art museum without stopping to gaze. This
perception requires an exercise of will. Ultimately, though,
only the effects of participatory perception in a person's life
can bear witness to its reality.

Participatory perception must be distinguished from
subjectivism. The color of the light of a Red Giant star, for
instance, can be measured objectively by a spectrometer.
Red then means that the spectral band of red light is measured as a higher intensity than that of the blue band. The
ratio of red to blue defines an objective color and is the same
for everyone who reads the display of the measuring instrument. Computer-controlled telescopes observing stars can
determine their objective colors automatically. Astronomers base their theories on them. On the other hand, there
is a subjective sensation of color that depends on the person
and the circumstances. The human eye sees faint stars only
with rod cells that cannot discern colors. For this reason we
perceive distant Red Giants as white dots. Only the closest ones, such as Aldebaran in the constellation of Taurus
and Arcturus in Boötes, may be recognized under good

observing conditions as being slightly yellowish. Other receptor cells in the human eye, the cone cells that can distinguish colors, come into play. The color becomes significantly more noticeable with binoculars. In color photos Red Giants finally become really red because the true color contrast between red and blue is enhanced. The subjectively perceived color is the reaction to an objective fact, which is dependent on the circumstances of the observing person. Pure subjectivism in this case would allege a color without reference to other viewers or the circumstances under which colors were observed.

The definition of science (from the Latin *definire*, meaning "to delimit") does not extend across the whole horizon of human experience. Some humans are deeply moved by a starry night sky, a piece of art, or a mystical experience. Such perceptions depend on the person and the circumstances, and are not reproducible as required by scientific methodology. Even so, I consider them real if they act to shape a person's life, feeling, and thought.

7.4 Depth of Reality

"As above, so below; as below, so above" was the motto of an old Egyptian mythology.[4] It claimed a correspondence between the divine and the earthly. The spatial concepts identified with "above" and "below" have changed, however, in our modern worldview. Today astronomers find that molecules in space radiate in the same way they do in terrestrial laboratories. The physical laws in the universe and on Earth are not just analogous, but identical.

Yet today we still perceive reality on different planes. They no longer have the former attributes of "above" and "below," but of "interior" and "exterior." Science explores the exterior reality by objective measurements and observations. Reality confronts us, however, also through participatory perceptions. Thus, fundamentally different perceptions are the basis of science, on the one hand, and religion and art, on the other. These distinct perceptions likewise span different planes of language and methodology. If the two planes of perception are not distinguished, misunderstandings and erroneous expectations result. Today, and for the past 400 years, such has often been the case in discussions involving science and theology. This fundamental difference is one reason why God will never be found by scientific methods, and why his existence cannot be proved or disproved by science. It is a hopeless endeavor to seek proof of God's existence or of a creation plan in scientific results. There is no direct path from scientific measurements to religious experience, since the two are enacted on different planes. Art, too, lies beyond the limits of the scientific method, which cannot take account of aesthetic experiences. Here we must recognize the limitations of every methodology. Some colleagues in the sciences are beset by a certain myopia, not recognizing the limits of their own disciplines and imagining the latter to be all-encompassing. It is at these very limits where the most revealing insights may emerge.

Even if science and those existential experiences upon which religion and art are based have different sources of perception, they can still relate to each other. An analogy in the form of "as interior, so exterior" may be invoked to describe certain parallel experiences in science and in human life. To begin with, an example of an analogy should be stressed involving the physical investigation of the smallest entities, where the tool of measurement is not sharply separable from the object. Both form a unity in quantum mechanical measurements in which the observer also plays a role. Since quantum mechanics is a fundamental part of physics, each scientific measurement has a participatory aspect.[5] Obviously, human subjectivity does not play an important role in this case, and the two planes should not be confused. We are speaking of an analogy, not congruence, between science and existential experience. The example is of special interest as it concerns a fundamental perception of reality in physics.

Another analogy between exterior and interior is the openness of the future. The future that approaches the cosmos, all creatures, and our human existence is not known at present. The long-term development of the galaxy and even of the solar system is as little known to us as our own future.

Understanding the multilayering of perception allows us to sense the depth of reality. Perception alone does not decipher reality; thus different perceptions do not imply a multitude of realities. We assimilate perceptions consciously and subconsciously until they become experiences and finally cognition. It is not possible, however, to articulate the deepest human cognitions directly, though we can hope to express them in metaphoric language from various perspectives, as is done in religion, art, and literature.

Science and theology are not competing theories about the formation of the universe or of the evolution of living creatures. My aim is to discuss in the following chapter the methods of interpretation and the difference between scientific explanations and theological statements, springing from the divergence in the perceptions of reality.

Figure 16: Barringer Crater near Flagstaff, Arizona, is 570 ft. deep. Here a nickel-iron meteorite fell 50,000 years ago that had a diameter of 150 feet and a mass of 300,000 tons. Its energy exceeded 70 Hiroshima bombs and devastated everything within a radius of ten miles. (Photo: U.S. Geological Survey)

8

From Perception to Interpretation

The stereotypical picture of an astronomer has him seated behind a telescope searching the depths of the universe. This is not quite reality. Astronomers don't spend most of their time observing, but rather analyzing and interpreting the data. An interpretation initiates new measurements. The scientific method consists of the interplay between experimentation and theory. In the course of this activity, different forms of interpretation come into play. To avoid any misunderstanding, let us consider three distinct kinds of interpretation in some detail: First, what the scientific method calls for and successfully does (*to explain and to model*); second, what scientists aspire to and discuss among themselves but don't publish in professional journals (*to comprehend*); and third, what they enjoy discussing with friends in the evening over a glass of wine at the fireplace, or what they write in popular science books (*to construe*).

8.1 Explaining and Modeling

What does it mean to scientifically explain a phenomenon? It must be pointed out that, in a scientific context, several explanations of complex phenomena are often possible. Each explanation is potentially wrong and requires confirmation. A nearly infinite number of tests can usually be imagined; and because it is impossible to conduct them all, an uncertainty always remains. Explanations may only be refuted by new observations. But while explanations that have become implausible are often adopted to new data, refined, combined, or complemented, they are rarely abandoned altogether so long as their originator remains on the scene.

Science understands some phenomena to be reliably explained, but other phenomena less so. The differences in reliability are immense, though it is useful to identify certain benchmarks within this span of continuity. The following terminological distinctions are commonly applied in astrophysics. *Well explained* describes a phenomenon on which there is general consensus. The number of "well-explained" phenomena has increased from a few tens in 1900 (explanations that are still valid) to some thousands today. There are no unanimous explanations because it is always possible to conceive and to test a new model. A general consensus does not guarantee truth. Thus, the theory of luminiferous aether, proposed in the seventeenth century to explain the propagation of light in a vacuum, turned out to be wrong after two centuries of consensus. It always causes a sensation when serious doubts are cast on a good explanation. Good explanations are supported by a meshwork of many observations and interconnect to other good explanations.

If a phenomenon is considered *unexplained*, it does not mean that no explanation is possible. There is always some preliminary hypothesis. "Unexplained" means that several explanations remain in contention, or that the most likely explanation has yet to be substantiated by further evidence.

Solar flares provide an example of how initially disparate explanations changed and finally converged to a single scenario—even though flares are still not well explained. A hundred years ago the speculation was that chemical explosions or avalanches created the flares. Today's explanations assume that magnetic energy is released. The phenomena in the solar atmosphere can be observed so well now that the physical explanations are becoming more and more reliable. Satellites capture X-ray pictures showing the thermal radiation of the two million degree corona, and how it is suddenly agitated by a flare and explodes. What was observed before at low-energy X-rays was like the smoke of a fire. More recently, the fire itself has been observed at high-energy X-rays—that is, the explosion center containing a gas of one billion degrees, with energy up to a billion billion (10^{18}) kilowatt hours being released within a few minutes. This energy amounts to ten thousand times the yearly energy production of humankind. New observations changed the explanation.

What do the observations show? The explosion is caused by local over-pressure. The extremely hot gas expands and blows out into space. There are interesting deviations during eruptions: the explosion often does not proceed far and gets stuck in the upper corona. Gravity can explain some of what has transpired. Yet there is an additional force at work, which constricts the flow of matter and causes loop-like structures. What is this baffling force? At this point science proceeds inductively: from the known to the unknown. Although it would be a lot more exciting, a scientist will therefore not posit the existence of a new force. Magnetic fields in corona are not directly observable, but known to dominate the corona; therefore they must be expected to be implicated.

That is by no means the end of the story. Since Galileo's time, physicists have not been content to propose qualitative descriptions but have aimed to reproduce observable phenomena through mathematics. It is the very heart of physical explanations. At this point, physics becomes difficult, so that nonmathematicians may feel uneasy. I ask the reader's patience, though, because what looks incomprehensible at first should become clear as the larger argument unfolds.

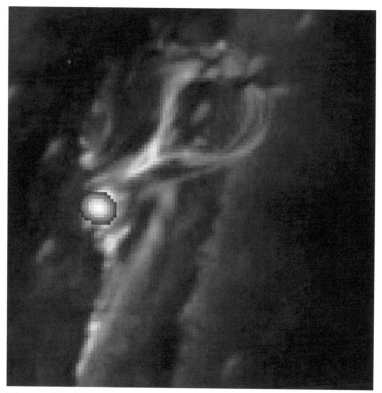

Figure 17: Flare at the solar limb (nearly vertical through middle). The image shows extreme ultraviolet emission of the hot corona and the ejecta of the flare observed by the EIT/SOHO space observatory (color-coded in blue). The extremely energetic electrons accelerated by the initial energy release emit hard X-rays (RHESS satellite, color-coded in red). (Photo: courtesy of Pascal Saint-Hilaire)

Physics always proceeds in three steps: First, phenomena are captured in quantitative measures (force, density, temperature, magnetic field strength, etc.). These concepts are averaged quantities and neglect microphysics. Second, the measures are brought into a mathematical relationship expressed as an equation. In our case it is Newton's second law,

$$\rho\delta V/\delta t + \rho(V\cdot\nabla)V \;\; = \;\; qE + (J\times B)/c + \rho g - \nabla\otimes p.$$

Incomprehensible? You do not need to make heads or tails of it. Expressed in words, the individual terms of the equation say that density times acceleration (on the left side) is equal (on the right side) to the electric force, plus the magnetic force, plus gravitation, minus the pressure gradient. The equation may be multiplied, recast, differentiated, integrated, and still agrees with the phenomena, as can be checked in nature. Mathematics is not only an encoded description but seems to be inherent in nature. I find this remarkable! The equation is valid even for processes in the Sun that are a billion times larger than in terrestrial laboratories.

The mathematical characters stand for physical measures and mathematical operations. They are abbreviations for concepts familiar to every physicist. Yet, the arrangement of mathematical symbols in equations is not identical to the use of metaphors in fiction. The difference becomes obvious in the verbal description of the equation above. The fact that it is possible to transcribe the equation into ordinary language might give the impression that the characters are nothing more than pictograms, where graphic symbols substitute words. This is not so. The verbal transcription does not reproduce the essence of the equation since the text does not contain the mathematics. Words cannot be multiplied or integrated. Because they mathematize nature, physics and other exact sciences describe nature on a different plane than do philosophers, theologians, and poets. To mathematize means to reconstruct quantitatively the observed reality using concepts and methods of mathematics.

In the third step, physicists finally solve the equation according to strict mathematical rules. The solutions to the equation have physical meaning: they describe how the density, velocity, and pressure change in time and space. They characterize the course of cause (forces) and effect (acceleration, eruption).

The above equation cannot be solved without information about the initial conditions. A spatial boundary and the physical parameters at the edge have to be known as well. Thus a scenario must be defined within which the course of events is described by the equation. The procedure is known as *modeling*. By modeling it, a phenomenon is *explained*. The model details, for instance, how energy of a magnetic field is transformed into heat and how hot gas explodes. Reality as perceived by measurements is reconstructed by a model. Science does not end here, however. The model must now be tested by further observations, and its correctness must be reviewed. The procedure of observing and modeling is cyclic. Models are not the truth; they simply must reproduce the observations. Most scientists are realists, however, and believe that models depict adequately a portion of truth.

To explain physically means to know the initial conditions, to find a causal relation between cause and effect, and to express it quantitatively by a mathematical equation. It is not always possible to establish a unique relation between cause and observed effect. Nonlinear phenomena such as the formation of stars and planets illustrate this. Such complex processes are described by deterministic equations, but their solutions strongly depend on uncertain initial conditions. Inversely, an observed phenomenon may have a large range of possible initial conditions. The causal relations in such a chaotic system cannot be reconstructed over a great span of time neither into the past nor into the future.

8.2 Comprehending

Only isolated processes may be fully explained physically and fully represented by a mathematical equation. In the universe at large, however, most processes involve a sequence of several causes and effects. When several processes occur simultaneously or in sequence, multiple equations come into play. They must be individually identified, so that the process can be reduced to simple subprocesses for analytic purposes. The reduction may be contrasted with the desire to overview the subprocesses in a larger scientific context and to *comprehend* the whole process in a nonmathematical way.

Comprehending means to view a phenomenon within the context of a larger scientific horizon and to reverse scientific reduction to some degree. The individual explanations then form a web of cause-and-effect relationships within a larger realm of reality. Scientists try to convey such coherent views to students in lectures and address them in review talks at conferences.

Solar flares are one example among many of complex phenomena in the universe where a plethora of individual processes are underway. They appear as details of the whole, but must be investigated one by one through the lenses of today's scientific subfields. The interrelation may be outlined as a chain of causes that are the effects of previous causes. Let me illustrate the causal chain of solar flares as an example of such complexity. In eruptions, magnetic energy is released. It is built up by thermal motions in the interior of the Sun and is in the form of electric currents. Intense currents in the dilute gas of the corona are unstable. The instability increases the electrical resistivity. Because of resistivity, magnetic fields can diffuse through the gas; it allows the field to simplify its geometry and to release energy. The important point here, beyond the details just mentioned, is this: cosmic reality is much more complex and multifaceted than simple test cases, such as planetary orbits, may suggest. The physical universe is no less dynamic than what we know of life on earth, whose complexity fascinates and frightens us occasionally.

The causal chain could be expanded with many more details; solar eruptions are a part of the larger chain that plays a role in the formation and evolution of stars. There is also a chain of effects becoming the cause of further phenomena. For example, flares accelerate particles, radiating in many wavelengths, from radio waves to gamma rays. Accelerated particles collide with gas particles and heat the lower layers of the solar atmosphere to millions of degrees. The hot gas expands, destabilizes magnetic fields, and is ejected from the Sun. Energetic particles reach interplanetary space. Ejections produce shock waves in the solar wind. In turn, shock waves accelerate more particles to nearly the speed of light. High-energy particles reach Earth, damage satellites, ionize the upper atmosphere, and disrupt navigation and shortwave communication.

One need not understand all of these details in order to understand the essentials of flares. What is essential is the energy available from the magnetic field and the complex dynamics of a hot gas or plasma that consists of electrically charged particles. Energy and complexity are the concepts that encompass the individual processes and allow comprehension. Comprehending means to distinguish the important from the less important, and to be able to develop and follow a relevant research strategy. In view of the complexity that confronts us in astrophysics, comprehending interrelationships may be as important as explaining individual phenomena.

Comprehending is a concept drawn from the humanities and should not be applied carelessly in scientific methodology. Wilhelm Dilthey (1833–1911) sharply differentiated between explanatory science and comprehension in the humanities. In the study of literature, for example, the point is to understand a text. It is necessary for comprehension that the investigator dismisses the objectivity of distance and put himself into the place of the author and his subjective perceptions. Whether it is possible to comprehend without explanations, or vice versa, is however disputed.[1] In particular, explanations of origin and historic context have contributed much to the project of comprehending ancient texts in theology. What scientific comprehension does have in common with comprehension in the humanities is that in both cases all collected and verified facts and explanations must be reviewed and assessed. The assessment separates the important from the unimportant and brings order to complexity.

We have started with an individual phenomenon, solar flares, and have encountered causal interdependences with an ever larger portion of the universe. One might then ask whether all parts of the universe are interrelated. Will we one day comprehend the universe? A promising beginning to such comprehension draws upon the perspective of temporality.[2] Time appears in physics both in causality and chance; both require a before and an after. Time, and not space, is the dominant dimension in astrophysics: all has formed; all will decay. Such comprehension is very abstract. And yet, it provides a structure to all knowledge. I consider it more relevant than single discoveries and explanations. In the second half of the twentieth century, astronomy began for the first time to comprehend cosmic interrelations, thereby revealing that the universe has a history.

8.3 Construing

The remarkable progress seen in astronomy raises this question: How is it that physics can capture cosmic processes in mathematical equations? This characteristic implies that a few basic forces are at the base of a nearly infinite number of unimaginably complex phenomena. Does this characteristic trait allow us to speak of the universe as a single entity, and what would doing so imply about the universe? For instance, does the universe work like one huge machine?

Answers to these questions constitute a special kind of interpretation. They are construed from experiences in other fields—such as mechanical engineering—and suggest analogies instead of mathematics. Such interpretation does not involve explaining through causal relations because these questions do not address cause and effect; the answers, therefore, cannot be mathematical. The questions on interpretation in the previous paragraph exceed the range of science and view it from outside. Such interpretations take place on a different plane. A displacement of the investigator away from the office or observatory to the fireplace, for example, is appropriate.

Scientists also pose questions about interpretations beyond the boundaries of science. In the course of scientific explanation, a scientist may experience a sense of wonder and ask where a phenomenon belongs within a universal context. Only the whole, and the relevance of the phenomenon to the whole, reveals its value. These construed interpretations are not published in scientific journals but appear in popular science books.

It may be argued that the universe does not need to be interpreted by metaphors since the questions being asked are not of the practical type that may be answered by science. They are not, in fact, relevant to astronomy. A possible stance, then, would be to assert that the universe defies interpretation. Interpretation, on the other hand, makes it possible to ask what meaning human existence might have. The question of the universe's essence touches on nothing less than the basis of our own lives. It is ultimately the question of why we are here, and where our place is in the whole.

8.4 Many Possible Interpretations

The Age of Enlightenment provided an interpretation of the world as a *clockwork*. In this mechanistic interpretation, cogwheels, a partial aspect of reality underscored during the emergent age of machines, become emblematic of the whole. The turning cogwheels reflect a pattern that was recognized in the universe, particularly as evidenced by the orbiting of planets around the Sun. The mechanics of the clock became the metaphor[3] for causal action in general. In this mechanistic metaphor the world is construed. It indicates that, on the one hand, the universe (including humans) amounts to a set of causally connected parts. On the other hand, it gives a value and meaning to each part. A clock would not work without each constituent component.

Any interpretation involves a temporary perspective. Themes prevalent within the science of a given era often provide the paradigm for interpretation. In contemporary terms, a solar flare, as described above, is not reminiscent of clockwork. On the contrary, it is a strongly nonlinear phenomenon, in which many parts interact simultaneously, making exact predictions impossible. The orbits of asteroids and the long-term orbits of planets are other examples. The chaotic motion of such nonlinear systems was not known in the eighteenth century. Quantum mechanical uncertainty and resulting randomness also don't fit the paradigm of a clock. The clockwork interpretation provides little enlightening in the present view of science.[4]

Today the interpretation of a *rational* universe is much in favor. This somewhat more general principle suggests that in principle the whole of reality is accessible through reason, is measurable, and may in principle be expressed mathematically even if science has not discovered all laws or may never discover them all. The rational interpretation is an extrapolation of scientific knowledge and projects current explanations—from the development of the universe to processes in the brain—to the whole of reality. Chance, which cannot be explained causally, must also be accepted within this paradigm. It plays a role in quantum mechanics, as well as in many other processes, including evolutionary biology. As long as it is pure chance, it can be handled mathematically. A mechanical clockwork no longer serves as the paradigm of a rational interpretation. A more relevant image would be that of a cosmic supercomputer with a perfect random number generator.

A rational understanding of nature and God constituted the basis of seventeenth- and eighteenth-century deism. Deists assumed that God created the universe and put it in operation like a clock, but exerted no influence thereafter. This rational interpretation excludes a God who intervenes deliberately in the processes of the world. There is only room for him in miracles that cannot be explained causally or by pure chance. All other processes are natural and do not require God. He may be imagined as an "Intelligent Designer," who applies rationality and goals at the outset but otherwise remains aloof from his creation.

Science assumes that the world is rational and can thus be captured in mathematical equations. This is not a dogmatic claim, but a hypothetical presupposition that began with Galileo Galilei's (1564–1642) study of free fall. It is part of science's method but also limits science's range of validity. Presupposing the world's mathematical nature has, as a working hypothesis, proven its value in the sciences for some four hundred years. Yet a working hypothesis is not an interpretation. It becomes an interpretation only if one postulates that the very basis of reality conforms wholly with mathematical laws. As a consequence, all reality in the universe is construed as a play of chance and necessity.[5] The retroactive interpretation has found much acceptance among scientists, but is open to criticism. Scientists certainly do not experience and interpret their lives and professional motivations solely in a rational way. The lack of completeness becomes most obvious in discussions on ethical issues; ethics cannot be derived purely rationally from chance and necessity.

The rational interpretation may also be criticized in that no system of logic can be derived completely from its axioms. The incompleteness theorem[6] of the Austrian logician Kurt Gödel (1906–1978) states that there are always assertions that can neither be proven nor disproven. Take, for example, the sentence: "I am not telling the truth." If the sentence is true, then its assertion is not and vice versa. According to Gödel, every piece of mathematical logic allows for contradictory assertions and is thus incomplete. Common sense may suggest that one should believe a person who voices the statement above although his words say the opposite. Thus, there are true assertions that make no mathematical sense and cannot be proven. The attempt to construe reality as altogether rational eventually confronts us with limits that are dictated by logic.

Another possible interpretation would suggest that reality is larger than the rational range of science, and that we therefore have to face a *suprarational* reality. Such a view acknowledges the existence of phenomena in the universe that cannot be presented in mathematical form. It insists that there will always be mysteries. Life and universe have a depth that cannot be completely revealed by science. The rational aspect of things that science explains then becomes a subset of reality. Such a formulation may be disconcerting, since it allows for all kinds of esotericism and occultism. Yet the poetic arts amply demonstrate that it is possible to deal competently with suprarationality.

One variant of the suprarational approach is an interpretation whereby the universe is seen to be embedded in an all-encompassing, transcendent reality. What, in other words, might it mean to construe the universe as a *divine creation*? Such an interpretation does not emerge from scientific explanations but construes them from a religious perspective. The role of the creator and his acting in nature has seen many metaphors: the prime mover, sovereign ruler, clockmaker, and sufferer. In a world dominated by a scientific worldview, his role may also be depicted as a creative artist, preserver of the cosmic processes, or group leader presiding over different powers of the world.[7]

Creation theology, better referred to as theology of evolution,[8] envisions new entities that appear spontaneously and contingently in the universe. A development is termed contingent in philosophy if it is possible but not necessary. The term is not synonymous with randomness, which is a property that may be described exactly in mathematical statistics. A contingent action may be deliberate and purposeful, but may not follow either law or chance. The word "contingent" originates from existential experience. Everyone has experienced contingent events in his or her own life: deciding to marry, conceiving a child, choosing a career, etc. Science declines to recognize the contingent emergence of new entities and attempts to explain such entities as the products of general laws or pure chance. For the most part, theology of evolution interprets even regularity in the universe as the product not of timeless laws but of contingent occurrences. It is God's will that creates and maintains regularity.[9] The presumably self-evident, regular state of the universe simply does not exist. The creation interpretation appears counterintuitive to a purely scientific worldview. He who talks about creation interprets scientific results in the light of other experiences. Creation interpretation does not aim at particular gaps in scientific understanding; it rather aims to comprehend the foundation of reality upon which all interpretation builds.

We cannot presume to draw on any generalized, consensus interpretation, although construed interpretations must face rational arguments. Interpretations are constructed by humans from their own individual or traditional perspectives and contain an element of subjectivity. Scientists may also construe the universe. Richard Feynman once said: "Just because we cannot measure position and momentum precisely does not *a priori* mean that we cannot talk about them. It only means that we need not talk about them. . . . The concept of an idea which cannot be measured or cannot be referred directly to experiment may or may not be useful."[10] But scientists do not always construe self-consciously and critically. So the interpretive aspect of one's outlook or general approach to things often goes unrecognized. Although interpretations may slip quietly into our discourse, taking the form of generally used paradigms, it is better to reflect critically on their identity—a practice that should not be foreign for scientists.

Surely, any number of possible interpretations exist. There are no bounds to how reality might be construed. And nobody, neither theologian nor scientist, has a monopoly on interpretation. This circumstance makes interpretations relative but not irrelevant. The available wealth of interpretations suggests that no one interpretation represents reality as it is. Yet interpretation amounts to more than icing on the cake of scientific explanation. The way we construe reality has a considerable influence on our behavior and outlook in life. Without interpretation, we cannot make reality accessible and cannot orient ourselves. Even to function in everyday life requires a measure of interpretation.[11] Yet our chief motive for interpretation would seem to be the existential desire to understand our experience rather than practical function alone.[12]

The sheer plenitude of interpretations does not forbid judgment. There are, for example, interpretations that bypass the realities of life and suggest either a frivolous optimism or a destructively pessimistic outlook on the world. Our motto might be phrased thus: He who construes things wrongly will be punished by life.

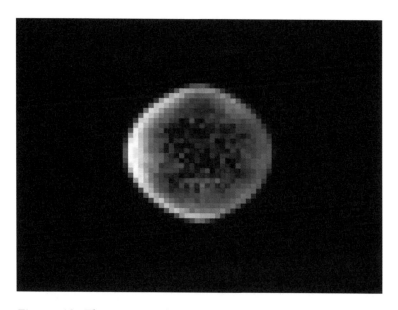

Figure 18: The picture shows gamma rays emitted by the Earth's atmosphere observed over seven years by the Compton Observatory satellite. Gamma radiation originates from cosmic particles accelerated in supernovae. When such particles penetrate the Earth's atmosphere, they collide with molecules of air and knock out electrons. Colliding with air molecules, electrons emit gamma rays. (Photo: D. Perry, NASA)

Figure 19: The two hundred billion stars of the galaxy NGC 4526 are visible in the picture taken by the Hubble Telescope only as a diffuse haze of light with a diameter of 70,000 light-years. The stellar radiation is intercepted by dark molecular clouds, the dust of which absorbs the light. In the lower left a supernova is noticeable. Its light is emitted by the ejecta of an exploding star. (Photo: High-Z Supernova Research Team, NASA)

PART III

Interpretation as Creation

The method of construing allows us to think further than science can, to the point of reconsidering creation. In this final part of the book, I do not want to judge the various possible construed interpretations, but to focus on one particular interpretation. It is a personal challenge to me as a scientist to understand rationally the radically different view of reality of the creation interpretation and to try to make it comprehensible to a critical reader.

What does "creation" mean—or can it mean at all—in a universe where stars form by causal laws and where there are no persistent gaps and obvious design? Can the notion of "creation" include the recognition that everything will also decay? Is it reconcilable with a dog-eat-dog world, in which the brutal necessity to eat or to be eaten seems to prevail? The faith-grounded concept of creation has always had to take account of worldly experience.[1] I want to examine the rationale, underlying assumptions, and possible consequences of interpreting the universe as a divine creation. Most of all, it remains to be shown that such an interpretation is not arbitrary or antithetical to the interpretive perspective of today's science but can logically coexist with it.

Figure 20: The atmosphere of Titan, the largest moon of Saturn, contains methane, ethane, and other hydrocarbons. They rain occasionally as fierce precipitation events and form lakes on the surface. In the radar image from the Cassini probe shown here, lakes appear dark and boggy coast regions gray. (Photo: Cassini Radar Mapper, JPL, ESA, NASA)

9

What Could Creation Mean Today?

In the Prologue at the start of this book, I describe my fundamental feelings of awe that made me decide to study astrophysics. What have they to do with the interpretation of the universe as a divine creation? In this final part of the book, I do not want to trade off the various possible metaphoric interpretations against each other, but to focus on this particular interpretation. It is a personal challenge to me as a scientist to understand rationally the radically different view of reality of the interpretation of creation and to try to make it comprehensible to me and to a critical reader.

What does "creation" mean—or can it mean at all—in a universe where stars form by chance and causal laws and where there are no persistent gaps of explanation? Can the notion of "creation" include the recognition that everything will also decay? Is it reconcilable with a dog-eat-dog world, in which the brutal necessity to eat or to be eaten seems to prevail? I want to examine the rationale, underlying assumptions, and possible consequences of interpreting the universe as a creation. Most of all, it remains to be shown that such an interpretation is not arbitrary or antithetical to the interpretive perspective of today's science but can logically coexist with it.

Long before modern sciences started to perceive and explain the universe, people construed what they learned from gazing at the sky and interpreted the cosmos as a divine creation. To explain scientifically, to comprehend, and to interpret metaphorically were not carried out in this chronological order. The arguments for interpreting the cosmos as a creation do not follow rationally from science but from basic experiences of participation such as described in the Prologue. Yet the faith-grounded concept of creation has always had to take account of worldly experience.[1] We cannot ignore today's scientific results when we ask about the meaning of creation. The new worldview does influence our interpretation.

9.1 The Creative Principle

Where and how might the concept of "divine creation" figure in present-day language? Amid the grand creativity of cosmic evolution, astronomical methods cannot detect direct actions of God. The objective approach of this science excludes perceptions, in which the observer participates, and thus excludes as well questions about transcendence or the researcher's existence. In the reach of science we can expect only impulses or consonance for concepts of creation but no proofs: telescopes observe physical processes and measure their parameters. Numerical models then yield plausible explanations and inspire new observations that can test these models. This part of our explorations, including the gaps we do not yet understand, belongs fully in the realm of the exact sciences. Nevertheless, every curious human will someday ask what is behind it all. Why, for instance, did the circumstances emerge that allowed interstellar gas to form stars? Why is it possible that such a complicated development culminating in the formation of life and in human consciousness occurred, and why is it possible at all that completely new things could form in the universe? Such questions deal with "rational origin,"[2] as opposed to astrophysical investigation of origin in space and time and by causality. Reason unites with the sense of wonder when we become aware how new celestial bodies form: the universe obviously has this astounding property of enabling new entities to emerge. Material structures do not remain as they are; they eventually decay but are capable of forming a new order.

There is a "Creative Principle," a fundamental property of matter that also shows up in chemistry and biology, as well as in human society. The principle is encapsulated in the common observation that the universe doesn't just change but rather shows us that in the course of change new entities form that had never existed before. The formation of new things implies the decay and passing away of existing entities.

Such emergence of the new in the universe should not be understood as the intervention of a supernatural force that displaces causality and chance. The Creative Principle stands in opposition to the idea of a "cosmic dance," of an endless cycle of formation and destruction.[3] The universe does not reflect a steady state of eternal cycling because the new is different from what preceded it, although emergent from the old. The principle is manifest in the evolution of the universe over the past 13.8 billion years. Today's stars are different from the earlier ones. There was a qualitative development from old to new: ashes of the first stars reappeared in the subsequent generations of stars and produced the novelty of the first planet. Thus, every new star and planet has its own composition and history. The same is true for humans. Although a human body contains many kinds of fusion products from the past, such as carbon and oxygen from burned-out stars, human beings are different from these stars, and each newborn child is a surprising novelty.

By the term "creation" some theologians mean "that

God creates something from nothing."[4] Thus, the really new, the creation, is not material at all. It is rather like the idea for a new picture that a painter may have in a dream at night. During the day, she then brings substance to the painting with brush and paint. As she works, the idea grows further, and only during the course of the day does the idea become a painting. The brush is the mediating agent between the idea that originated in the painter's brain and the paint that was produced in a chemical factory. The example demonstrates the point that the new has several origins, depending on the plane of reality in view.

The new entity in question forms from existing substance but cannot be regarded as a straightforward derivation from the old. It contains old material, but is subject to a new order. The word "new" suggests that it formed in a spontaneous way. Often it organizes itself through the strengthening of a small deviation, such as a slight increase in the gas density of an interstellar molecular cloud developing into a cloud core. We have already noted how incredibly creative various processes in the universe are. There seems to be an inherent principle that continuously initiates new developments in the universe. Science approaches this principle on various levels and different fields. In the context of thermal fluctuations, it has been described by the Belgian chemist Ilya Progogine (1917–2003) as bringing "order out of chaos."[5]

The Creative Principle is a cosmic property as opposed to the concept of divine creation, which relates to this creativity and interprets it. But let us first explore the relevance of this principle to the formation of life.

9.2 The Formation of Life in the Universe

When I gaze at the sky at night, I sometimes feel as if I'm sitting in the gallery of a theater watching a dramatic performance in a barren universe. From this viewpoint it is utterly astounding that something as complex and subtle as living creatures exists on Earth. How could they have developed here? Do such creatures form in many places, or is the Earth unique? The formation of life is an essential ingredient of any notion of creation.

Life as we know it can form only when its essential components, such as carbon, nitrogen, oxygen, and other atoms, are at hand. They are necessary constituents of organic molecules that are the basis of life. The atomic elements originated in the interiors of stars and were carried by stellar winds and supernova ejecta to the interstellar gas, out of which new stars and planets formed. Life is only possible thanks to previous stars. Every fiber of our body contains ashes of at least three generations of stars. The ground on which we stand is stardust as well since rocky planets like Earth could not form without previous stars of high mass. They fused huge amounts of hydrogen to iron and nickel that make the core of Earth. Stars are an indispensable link in the chain that leads to planets and life.

More than half of the approximately 150 molecules known today in molecular clouds and stellar envelopes are organic molecules. The carbon chemistry (known as organic chemistry) in space forms molecules consisting of up to more than a dozen atoms. In meteorites even larger molecules are detected, including amino acids, the building blocks of proteins. They demonstrate how rich in organic molecules the solar nebula and the planetesimals were. The Earth formed out of planetesimals and their debris. Most of their organic cargo was destroyed in the early phase of planet formation, when the temperatures exceeded the melting point of rock. Later impacts of comets and asteroids have undoubtedly deposited anew simple organic molecules on the surface of Earth. Nevertheless, it would be a bold hypothesis to suggest that life originated in space, as considerably more complex molecules are necessary for life than are found there. These are very fragile, and thousands of different kinds are needed. Yet life may in fact have developed from simple organic molecules that originated in space. The oldest rocks on Earth demonstrate that life formed soon after the Late Heavy Bombardment, and not later than three-and-a-half billion years ago.

Large asteroids that impact other planets fling pieces of rock into space. These can reach Earth and are found here as meteorites. Rocks from Mars may be identified by their unusual mixture of argon isotopes in gaseous inclusions. A Martian meteorite found in 1984 in Antarctica contains microscopic structures surprisingly similar to those of single-celled fossils on Earth. The "discovery" of life on Mars was highly controversial and ultimately dismissed, but it motivated NASA to engage in a series of Mars missions with the goal of searching for life or past biological activity there. In addition, the Mars meteorite suggested a way that life might propagate in a planetary system.

For half a century there has been a deliberate search for life in the universe. In 1960 the American astronomer Frank Drake was the first to monitor the two nearby stars *Tau Ceti* and *Epsilon Eridani* using a radio telescope of 80 feet in diameter. All attempts to search for extraterrestrial intelligence (SETI) failed, however. Another path to searching for life in the universe leads to the planets of the solar system. The Mars missions involving the two Viking spacecraft in 1976 took soil samples and searched for biological traces, but with no success. Other places in the solar system where our kind of life is considered possible and which may become targets of future missions are Jupiter's moon Europa, under whose ice-covered surface there is liquid water, and Saturn's moon Titan, which has an atmosphere rich in methane and hydrocarbon molecules.

To decide where to search for life in the universe, assumptions must be made as to how life forms. Life on Earth seems to have originated in liquid water, a setting in which organic molecules dissolve easily and react with one another. Some science fiction novels speculate about silicon-based life. Silicon, unlike carbon, is however not a component of air since it combines with oxygen to form a solid compound at temperatures below two thousand degrees; this compound does not dissolve in water. Though we may simply lack the imagination to conceive of alternative systems, it seems that life has good reasons to base itself upon organic chemistry. Based on this assumption, several prerequisites must be met for life to form, and liquid water is generally accepted to be the most important. To have liquid water, a planet must orbit its central star in a near-circular orbit and at a certain distance similar to that of the Earth to the Sun. It took three billion years for multicellular life to form from single-celled protozoans. During this time the planet could not be hit by an asteroid larger than a hundred miles in diameter, or else its oceans would have evaporated into space. The central star had to be stable during this time. The latter condition is only fulfilled by stars whose mass does not significantly exceed that of the Sun.

In 2007 the first planet other than Earth having some of these properties was found: Gliese 581c.[6] It circles a star 20.4 light-years from us that has a mass one-third that of the Sun. Low-mass stars are less luminous, but this planet is also comparatively close to the star. The average temperature on the planet is between 27 and 104 degrees Fahrenheit. Due to the small orbital radius, the planet's rotation period is synchronous with the orbital period. Thus the same side of the planet faces the star continuously, and that side heats up. On the dark side, however, water is permanently frozen. The local temperatures deviate greatly from the average value. It is therefore questionable whether the planet has much liquid water and complies with the many other conditions necessary to form life. Currently, the search for Earth-like planets is being pursued quite eagerly. Several future space missions will not only detect many Earth-like planets, but even observe them directly, in spite of the glaring light of the nearby central star.

How could life be detected on another planet? The first living creatures on Earth were cyanobacteria that lived in water and carried on photosynthesis. They produced organic compounds from sunlight, water, and carbon dioxide while releasing oxygen. Oxygen reacts rapidly with other elements on the surface of a planet, including silicon, aluminum, and iron. If oxygen is not supplied continuously by photosynthesizing living creatures, it rapidly disappears from a planet's atmosphere. A high abundance of oxygen molecules (O_2 and O_3) in the atmosphere and liquid water at the surface would be indicators of life. Both may be detected by their characteristic spectral lines. It will be exciting to apply this "oxygen test" to Earth-like planets.

An enormous number of Earth-like planets probably exist; the latest results suggest that there are more planets than stars in the galaxy. Recent observations of exoplanets clearly show, however, that a large majority of them is incapable of fostering life. Thus, we do not know how abundant life is in the universe. But we do know that single-celled creatures existed on Earth just a few hundred million years after the Late Heavy Bombardment, thus relatively soon after the time when life on Earth became possible. This fact is often taken as evidence for the abundance of single-celled life elsewhere in the universe. And we do know that its plenitude and diversity on planet Earth are simply amazing and demonstrate a stunning creativity.

9.3 Anyone Else out There?

It is not known in detail how atomic building blocks formed single-celled creatures like the bacteria and how these then developed into multicellular animals. For sure, the latter took at least five times longer to develop. Multicellular life unfolded probably after a major but not total catastrophe. The latter development appears to be a stroke of luck that may well not have happened at all. The frequency of multicellular life is the product of a tiny probability within a huge number of planets. The conditions necessary for the emergence of multicellular life are not sufficiently known, so the question about how many such planets there are remains highly disputed. Even less can we presume to say how many intelligent civilizations there may be throughout the universe. On the other hand, so far as we know today, there is no manifest reason why the existence of life beyond the Earth should be considered impossible.

And what if they exist, these extraterrestrials? Let us conduct a thought experiment and assume that there is intelligent, extraterrestrial life and that we will meet it one day. We should certainly find it intriguing to exchange scientific knowledge, in the way that it is so extensively described in science fiction novels. Could we also look forward to broader forms of cultural exchange? Would these aliens also show an interest in art and religion? Would they interpret their existence as the product of chance or of divine providence? Our speculations on such matters depend largely on how we interpret our own existence and on our perceptions of reality.

The goal of this interplanetary encounter should be some form of authentically mutual cultural exchange. How, for instance, could one communicate to extraterrestrials a creation narrative? We should wish to convey not only the framework of the story, with its particular worldview, but its actual content, including what we take to be the order and value of all things in the universe, and our sense of life's meaning. How could one translate such matters in a way that makes them understandable and plausible to extraterrestrials? Do they have a sensorium for participatory perceptions and religious experiences? The question forces us to wonder once more about the character of religion, and about how it relates to reality. For sure, prospects for our enjoying peaceful relations with extraterrestrials could depend quite critically on our ability to share with them something of earth's longstanding cultural and religious values. For this reason, cultural exchange may be more important than communication about scientific and technical matters.

So much for the thought experiment about extraterrestrials. Perhaps its significance lies not in the astronomical realm so much as in what it can suggest to us about a general deficiency in contemporary culture. Consider, by way of analogy, a simple rule applicable to most university teaching: the better a lecturer has come to understand a difficult subject, the better his or her students are apt to learn it. Deep familiarity with the subject is contagious. Quite a number of our contemporaries, though, find it troublesome to understand traditional concepts of creation and have not been acclimatized to inquire about its fundamental experiences and ethical implications. Our civilization lacks not only a more-than-superficial familiarity with religious ways of thinking and feeling, but also a vision of how these values might be translated into a language consonant with today's worldview as shaped by science.

9.4 Ethical Reflections

A human being, considered in strictly spatial and temporal dimensions, represents only a minute portion of cosmic space and history. If one focuses instead on system complexity, though, the human brain and its capacities look impressive, even on a cosmic scale. The functioning of a body on the molecular level of a multicellular creature is probably far more complex than the processes in a star. The complexity in question results from the multitude of significant mutual reactions of all elements involved and the large number of equations that would be necessary to describe the system. It is a characteristic property of the universe that such complexity can arise, although it will always be local and limited in space. We do not know, and may never know, whether biological evolution on Earth is a singular case. So for the following ethical considerations, we have to presume that biological development only rarely proceeds so far that multicellular or even self-conscious creatures form. There is no evidence to exclude the possibility that biological evolution on Earth extending all the way to intelligent life may even be unique in the whole universe. The possibility lends tiny Earth and humanity an extraordinary significance at the cosmic level.

More and more in the course of history, humanity also contributes to biological evolution on Earth. Humanity's impact has important consequences for terrestrial life; seventy species go extinct each day.[7] The current acceleration in species extinction is more than one hundred times greater than before modern times. It will cause a lasting break in the history of the Earth and, in view of life's possible uniqueness, within cosmic history as well. It is not yet foreseeable how large the break will be compared with previous catastrophes caused by cosmic events. From the standpoint of geophysics, evolution of life on Earth could continue, although with setbacks, for more than a billion years. The largest risk is humanity. How barren a universe this would be without the presence of conscious beings to admire the stars!

Ethical postulates cannot be derived directly from the framework of science. On the contrary, once we realize that life has developed through selfishness[8] and is stained with blood, we may not find it easy to develop an empathetic and moral attitude. The survival of humanity is certainly of supreme value. Neither the human species nor the environment, however, is immutable. Both have changed significantly within the past million years. Yet we need to reconcile our ethical outlook, within the capacious worldview of cosmic development, with the idea that the development of Earth that has proceeded now for more than four billion years must not come to a halt. Our actions must not be guided by the short-term interests of the human species. There must be a larger "Reverence for Development,"[9] an ethical principle to guide our life on this wonderful planet Earth. We take part in its development and bear responsibility for sustaining it. It is not possible here to present a fully developed code of ethics. Ethics can grow only from a consideration of all parties and circumstances that is balanced in all ways. In the following some aspects of this ethics are presented.

With respect to universal development, our duty is to nurture our own development without inhibiting other forms of development taking place around us. Reverence for Development may also include a concern to pass life on to descendants. The successors may include biological offspring or the intellectual education of future generations. This ethic displays solidarity with the development of all creatures, not only of our own species, and is in accord with the Earth's development.

We do not live in a closed system that must be conserved at any price. The universe is not a machine but a developing system. As long as the development is linear, moving along the expected track, it can be predicted. That will not go on forever. After a certain time—in statistical mechanics it is the Lyapunov time (p. 84)—the development will take a new, unpredictable direction. We are not moving in a train but in an off-road vehicle capable of driving cross-country in several directions. But the journey could also end at some point. The development that produced such a magnificent plethora of life forms and landscapes may end in a barren desert. As a single human being I can only be amazed that the development is continuing, and I must do my best to support it.

Development on Earth has long ago broadened from biological evolution to include intellectual and cultural values. In fact, it is within the evolution of human societies that the largest changes have occurred in the past ten thousand years. Earth's diverse human cultures grow together, driven by the global technological civilization and its passion for progress. Some cultures disappear while others adjust and develop in the new environment. Cultures develop in close relation to their environment and must adapt when it changes or disappear. Such development is occurring presently, for example, reacting to the human threat to the environment. The cultural backlash is obvious: major disturbances of our natural habitat have spawned the vigorous present-day conservation movement—a remarkable development indeed.

Recognizing that development occurs on many fronts does not, in itself, give rise to ethical principles. On the contrary, many selection processes in biological evolution seem to contradict all moral rules. Yet, on the plane of human society, rules beyond the law of the jungle do and must apply. The ethics of the Sermon on the Mount do not demand that one has no enemies—just that one is obliged to love them. How refreshingly paradoxical! Furthermore, the exemplary life of Jesus demonstrates that he who loses may be the real winner. Thus Reverence for Development must include solidarity with underprivileged and unselected creatures.

Even if the evolution of creatures were to end tragically one day and life should disappear, the process of development deserves our admiration and awe. Reverence for Development means planting an apple tree "even if the world ceases to exist tomorrow," as Martin Luther put it.[10] Why? First, because apple trees have never been planted with the expectation that they will last for all eternity. We plant and hope, knowing well that apple trees may sometimes not survive one winter and may never even bear fruit. To plant an apple tree is symbolic of starting something new and our affirmation of development. If development has had meaning over thousands of years, then it retains its meaning this afternoon and tomorrow. Second, the world will not disappear into nothing. True, we cannot maintain the status quo forever. One day Earth will become uninhabitable, and life in its present form will be extinguished on this planet. Perhaps, though, something new and yet unimaginable will then form, and the apple tree may be part of it.

This view is based on hope. A certain expectation of future development is part of this hope. In seemingly desperate situations, hope focusing on a visionary future yields the strength to act ethically now.[11] Whatever the evil, hope carries the presumption that everything will bear some positive meaning at the end.

9.5 Hope in Spite of Decay

To hope for the future in a developing universe where all will decay can only mean to expect something new, something that does not yet exist or that can merely be hinted at. Expecting something new, we interpret the signs of the time hopefully. Hope is more than optimism, which flourishes only if one perceives good reasons for a favorable outcome. One may still hope in a forlorn situation and do so against reason.

There is a remarkable asymmetry between the decay of all things in the universe, a development that may in some cases be predicted accurately, and the formation of new things, which cannot be anticipated; we can only hope for the new. Of course this expectation may be an illusion, or a false hope that ignores the relevant facts. So how do we acquire the right hope?

Hope is based on experiences of rewarded trust. These experiences are not proofs or even predictions. Hope that focuses on the future, instead of factual evidence in the present, expects the new. Science cannot offer hope, since experiences of trust are not part of science. Neither does science exclude hope, however. Science assumes a regular state of the universe. For example, science expects that in the next instant, 400 quadrillion ($4 \cdot 10^{17}$) seconds after the Big Bang, a new second will arise. There is no proof of this. The historical development of the universe may serve, however, as an obvious paradigm for the future. Many new things may form, as the universe has been extremely creative up to this point. It is only a few hundred thousand years since human consciousness arose, a mere blink of the eye as compared with the age of the universe. Completely new things form these days in human society. Hope refers to such observed creativity.

Will new things really form in the future? The new cannot be predicted at all. It forms spontaneously and chaotically because the universe is not like a mechanical clockwork and does not evolve in a linear fashion. How should science foretell something that has never existed before yet has a complexity comparable to that of stars or living creatures? For this reason, the future development of the universe remains open to a considerable degree. The "open" aspect does not concern decay—which will occur eventually—but the formation of new structures from the old and the decaying.

Hope for the new is one of several patterns for interpreting the signs of the times. Those who adopt this pattern interpret current and future developments in a given way. Alternatively, one could adopt a pattern that leads to expectation of terminal decay, despair, or agnosticism. Once we settle on a mode of interpretation, the scientific facts begin to appear in a certain perspective and in a certain light.

The central message of Christianity to the world is hope, a hope in which even decay and death do not have the final word. The sequence of Good Friday and Easter is the prime example of this hope. From this perspective, hope expresses itself in the following way: the creator of the present universe will also create the world of the future. His benevolence will last even if all decays. Here the notion of creation has a component of the future and in itself signifies hope. Acting ethically may thus be seen in consonance with creation, reflecting gratitude for what one has already experienced and one expects from the course of developments.[12] Hope and trust yield the necessary strength for an ethic in which one becomes willing to face an open future.

My use here of terms such as "creator" and "benevolence" is meant to apply to a contemporary worldview, so that their signification differs from that conveyed in other historical contexts. What was once conceived to be self-evident language with respect to "God" as "creator" has become a minefield of misunderstandings in the present-day context of scientific discourse. It should first be noted that such concepts must be understood as figurative elements of a symbolic language. Originally they provided images for certain existential experiences and indeed had nothing to do with the universe at large. The source of these images is everyday life. In English "to create" is language closely associated with the notion of an artist who develops a new piece of art from raw material such as rock or paint. The word is akin to the Latin *crescere,* which means to grow. It combines the idea that new things are formed from pre-existing material with a recognition that completely new entities emerge from the hands of the artist.

To speak about the "creation of a human being" is also to use metaphoric language. It does not refer to biology but to the experience whereby a person becomes self-conscious, realizing that neither he nor his parents are the ultimate cause of his own being but that he received his existence ultimately from some other source. Such a person may then say, "God has created me." By contrast, the notion of "creator of the universe" tries to combine two concepts, each of which relates to a different plane: the above existential experience (with reference to "God") and the scientific term ("universe").

As noted before, the material things in today's universe did not originate in the Big Bang, but developed over the course of time. Currently some ten quintillion (10^{19}) stars are forming in the universe. The process of star formation takes some ten million years. Thus, on average, 30,000 stars complete their formation phase each second. In other words, in the present universe tens of thousands of new stars are born per second to become a potentially notable part of the universe. The modern notion of divine creation has to include them, which would suggest that God creates the very same stars that astrophysicists observe forming today. Can this make sense?

We will ponder this question in the following sections, where the aim is to see how the symbolism of creation, based on existential experience, might be brought into an enlightening relation with scientific views of the universe's formation and evolution.

10

God in the Universe

In Isaac Newton's chief work, the *Principia* (1687), the word "God" appears sixty-three times. During this era the concepts of God and universe were closely related. Today the word "God" is completely missing from scientific publications. It does not appear in any of the fifty thousand research articles in astrophysics published each year. Of course, the words "beauty," "meaning," and "love" are also missing. For two centuries now, science has avoided all reference to concepts of this sort and does not ask metaphysical questions. One reason why progress in science is so rapid is that issues to which no consensus may be expected are bypassed.

The notion of God varies among different religions, but even more importantly it varies with time. In antiquity, Greek popular belief imagined the gods as large humans, presumably about ten times as high as a mountain, thus making them about ten miles tall. The universe as known today is at least 10^{22} times larger than a mountain. Does this drastic shift in scale change our notion of God? I think it does. A Roman Catholic priest asked me after a presentation: "How can I lead a Corpus Christi procession out into God's nature if the universe is inconceivably large? Then God must be even larger." The number 10^{22} is so unimaginable that all notions of God become inadequate. And indeed the image ban in the Abrahamic religions forbids any picturing of God.

Books of various kinds on the issue of God and science are not in short supply. They often argue philosophically and take pride in treating this subject objectively. Yet religion involves the whole person, not only his or her intellect. Since God cannot be experienced objectively, religion cannot be objective and independent of the person. Thus God cannot be part of the reality that science is investigating, and science is as unlikely to find God as the scientists in Flew's original version of the Invisible Gardener (see page 63). There is no compelling proof of God in laboratories, in the formation processes of stars, or in the Big Bang. Nor does God appear in the mathematical equations that describe these processes. The notion of a "watchmaker" God from earlier times makes little sense in the context of our newfound awareness that the universe is still developing and that new entities are forming today. Though it is impressive that something as complex as human consciousness could emerge in the universe, the metaphor of a plan or "intelligent design" does not accord with the history of cosmic development, especially in light of what is known about monstrous cosmic catastrophes, dead ends, and the immensely extravagant aberrations in biological evolution. Science, and the worldview associated with it, is perforce agnostic. The scientific method of measurement and explanations of the data by cause or chance can reveal no evidence of supernatural invention.

And yet God, the Divine, or the Transcendence have not simply faded from public'sight, as have many other old notions, and the reasons for this lie surely outside the domain of science. It is true that earlier conceptions of God as construct for explaining scientific results, in the manner proposed by the physico-theologists around 1700, is no longer necessary. But perceptions such as those of Blaise Pascal (see page 98)—visions, or life-changing personal experiences—continue to constitute reasons why we must speak of God.

Yet if God is experienced only through human awareness, the question remains as to how the religious interpretation of the universe as a creation relates to the cosmic
reality observed by astrophysics. The authors of the biblical
creation stories posed a similar question, and it is illumi-
nating to consider their answers. Already during biblical
times the notion of God did not refer to results of objective
measurement and theory. The biblical reasons for speaking
of God had a different origin: God was experienced in life-
threatening danger or in catastrophes, as in the flight of the
Israelites from Egypt, or at the execution of Jesus.

10.1 Biblical Creation Stories

In Western culture, the two creation narratives of Genesis
1:1–2:4a and 2:4b-25 have traditionally claimed prominence.
What was generally understood at the time of their composition may still be identified today as the central message:
that the universe is embedded in something or someone all-
embracing. This transcendent force or presence willed the
universe into existence, bestowing upon it a causal unity
and final reason. The biblical creation stories differ from
that presented by some other ancient religions in that the
latter attributed divinity to the cosmic entities featured in
their narratives. Today we should add that modern science
treats not only the entities but also their formation process
as nondivine.

That the Hebrew Bible assigns human traits to the transcendent power should not be unsettling, so long as one
recognizes them as metaphors (from the Greek *metaphora*,
to carry across) based on human experience. Genesis 2:7
describes how God made man from dust of the ground. The
metaphoric coloring cannot be overlooked: the origin of
this image may be a potter who takes amorphous material
and gives it a form born of his own will and imagination.

The Genesis authors avoid use of the word "clay," alluding instead to fertile soil. The very name "Adam," bearing the generic sense of "human being," is akin to the Hebrew word for topsoil, *'adama*. According to modern biology, the human body does indeed consist of atoms from the soil. Yet a metaphor should not be validated by such an accident of superficial "truth," but by what it can convey and achieve. During biblical times, the distinction between the origin of the image (soil) and the image's referent (the human body) would have been clear to readers. The metaphor aims to say that a person forms from and derives continuing life from what the soil yields—and finally degrades to become soil once again. Yet soil provides only the material ground of this creation. It is only when God animates this form with breath that it becomes a human being. Ultimately, then, the metaphor expresses the concept that every person originates from a divine will, embodies a divine idea. So humans do not originate most essentially from soil but from an idea. In the history of Christian theology, the notion of origin from an idea has prevailed over that of origin from material only since the time of Augustine (around 400 CE). It is known as "creation from nothing."[1]

The metaphoric story of the creation of man in Genesis 2:7 expresses the existential insight that neither we nor our parents created us. This insight is the very origin of the general concept of creation. It is an existential experience—the discovery that *life is a gift*. It is to be treasured as such even when the creation narratives go on to acknowledge that human existence must be played out in an ambivalent world of labor, mortality, and adversity rather than in paradise. The experience of giftedness also suggests, of course, the presence of a giver, and gives rise to the metaphorical conception of a "personal" God. The extension of the creation concept to the universe is founded, in turn, upon this primary experience of creation: only those who understand themselves as created can also understand the universe as a creation.

Some creation stories that interpreted the universe metaphorically by its becoming before time were already difficult to understand in Greek antiquity. In contrast to previous narratives, early Christian witnesses and theologians of
the first century interpreted God's action in the world in
a startlingly new way. The novel outlook is based on the
extraordinary testimony—some of which is attributable
to eyewitnesses—recorded in many passages of the New
Testament. Focused on the aftermath of Jesus's crucifixion,
such testimony includes accounts of several episodes in
which Jesus appears before his disciples. A number of disciples, shocked and disillusioned by their master's execution, had already returned to their previous occupations as
fishermen. But the tide turns with the reappearance of the
one supposed to be permanently deceased. Something new
formed out of the catastrophe.

It is worth emphasizing that, according to the uniform
testimony of the New Testament writers, Jesus's resurrection signifies a change more decisive and cosmic than the
miraculous resuscitation of a corpse reported in other biblical episodes such as the raisings of Lazarus and the widow
of Naim's son. The resurrected Christ is portrayed not
simply as a revived dead man but as a prototypically new
and *transformed* human being. While his physical disposition is said to be new and different from anything before,
his past has not disappeared completely. His wounds
remain tangible. And yet, the resurrected Christ is described
as radically transfigured, walking through doors, moving
from place to place more swiftly than pedestrian locomo-
tion would allow, and being at times barely recognizable to
those who presumably knew him well. He is a new creation
come about in the present time.

Unlike most ancient views of postmortem survival,
including Greek ideas about immortality, the resurrection
of Jesus is indeed—contradicting the famous phrase from

Ecclesiastes—a "new thing under the sun." The apostle Paul's First Epistle to the Corinthians extends the narrative to affirm that not only Jesus but anyone who is "in Christ" is henceforth part of "a new creation; the old has passed away, behold, the new has come" (2 Cor. 5:17). For Paul, moreover, the new creation assumes significance to the cosmos at large, since by virtue of the resurrection "creation itself will be set free from its bondage to corruption" (Rom. 8:21).[2]

In fact, the resurrection narratives—including the accounts of Jesus's postexecution appearances by the Gospel writers and Paul's theological commentary on the resurrection event's relation to new creation—offer an interpretation of creation that is even more central for Christian faith than the creation stories in Genesis. Accordingly, the last book of the Bible, the apocalyptic book of Revelation, envisions an enthroned and glorified Christ who presides over "a new heaven and a new earth," declaring that "Behold, I am making all things new" (Rev. 21:1, 5).

The New Testament's resurrection narratives thus propose an unprecedented rationale for preserving hope in the face of crucifixion, death, and decay. A new community arises and celebrates. Easter represents the paradigm of how catastrophes may lead to something other than a definitive endpoint or annihilation; God can create new things out of the old. The Easter story became the cornerstone of Christian belief in an all-encompassing vision of creation. So, for example, in the Prologue to the Gospel of John, experience of this creative impulse becomes paradigmatic of an influence exercised upon the whole world:

> In the beginning was the Word, and the Word was with God, and the Word was God. . . . All things were made through him, and without him was not anything made that was made. . . . The true light, which enlightens everyone, was coming into the world. . . . And the Word became flesh and dwelt among us. (John 1:1-14)

The early Christians perceived in Jesus' speech and presence a power that shook and changed the world. They related this creative force (which John called "Logos," that is, in our translations, "the Word") to the formation of new things throughout the universe.[3] The words of Jesus are construed as forms of the divine wisdom that already existed at the beginning of the universe and emanated from God.

Granted, this idea is not easily grasped today. But I think we may better understand it if we envision John's prologue not merely as an introduction but as the abstract of a theory that is expanded and made apparent in the remainder of the Gospel's account of the life, death, and resurrection of Jesus. According to John, these elements of Jesus's story are exactly what unfold when God's creative force enters our world. Without this interpretive outlook, John could not comprehend the particular events he narrates in the following, or would not narrate them in the way he does. He relates them to cosmic creation and at the same time reveals what he means by creation with reference to the Word (a multivalent Greek word, suggestive of an ordering or formative principle as well as Wisdom) that originates in God and changes the world. Creation does not occur in a primordial, mythical era. John sees creation enacted through Wisdom's animation of cosmic evolution since the very beginning of things. But for John, creation is also manifest in latter-day history, above all in the case of Jesus of Nazareth. Participatory experiences in the life of the disciples are perceived to be metaphorically transferable to the cosmos at large: as interior, so exterior. In the process of this transfer, the universe becomes figuratively interpreted as creation, and at the same time the notion of creation is explained as analogous to the life and resurrection of Jesus: a development that includes suffering and annihilation but transcends them in something new.

Is such an outlook still understandable today? What would it mean to say that God creates the currently forming stars? Only participatory perceptions, rather than scientific findings or gaps in scientific explanations, can initiate the idea of a divine creation of stars. Such perceptions are often associated with a sense of wonder, as in Walt Whitman's poetic discovery of mystery and beauty in the starry heavens. Alternatively, the perception could be one of fright, inspired by the known hazards and finite nature of the solar system or the forbidding, inhuman immensity of outer space, as noted by Blaise Pascal (next chapter). Creation stories serve to convey meaning to such perceptions, disclose order, and help to orient individuals and whole cultures. The perceptions of creation are beyond the range of scientific measurement and observation. They cannot be expressed mathematically and are not objectively measurable to make creation plausible without some form of human participation. To comprehend stars as creation there is no other way than to experience by oneself, to step outside on a clear dark night and to look at the stars.

Stars and their physical processes can be perceived as creation only if viewed not simply as the inevitable norm, but as remarkable gifts. The formation and evolution of stars are far beyond the abilities of humans to influence. We cannot produce stars, but also cannot live without them. Humankind could not have come into existence without the previous existence of stellar bodies, and is completely dependent on the nearest one, the Sun. A sense of wonder at having been given our lives, along with the nurturing universe we inhabit, has been expressed in numerous biblical creation stories and psalms. It is possible even today to understand and identify with this sense of creation when reflecting, for example, on the way that new planets form from the ashes of old stars. Such new formations offer cause for hope since they are occurring in our own time. To believe in creation is to know that the evolution of the universe continues to unfold and is in good hands.

10.2 Cosmic Icons

On one occasion, after I sat for a television interview about the development of the universe and creation, the stage director told me that she found the film clip engaging and the pictures fascinating, but that she had difficulty concentrating on the interview. From the first few minutes of the interview she became preoccupied with the question of why I was talking about God at all and whether he even existed. After all, I had not identified any traces of him, nor had I even given any hints that would argue for his existence.

Can we locate any traces of God? Yes and no. First the "no." It is not at all a given that stars and planets should necessarily be able to form in the universe. Even less certain is the formation of life. Our universe would be completely different if some parameters in the standard model of particle physics were different from the measured values. If electrons, for example, had the mass of their sister particles, the myons, neutrons could not decay. There would only be neutrons and neutrinos in the universe, no atoms as we know them and no molecules or human beings. There are well over a dozen requisite physical constants that make life possible. The long list of requirements is known as cosmic fine-tuning.[4] It is unexplained, surprising, and wonderful, but not a definite trace or proof of the existence of God. An almighty creator could also be expected to create humanity in a nontuned universe, or by fixing the parameters locally to the values necessary for life—perhaps just in a particular galaxy or even on a single planet. That particularity would indeed be a physical trace! The fine-tuned parameters show again that God's existence cannot be proven by scientific methods. What is true for fine-tuning of the universe holds also for the finely threaded chain of processes involved in star formation and for the unbelievably favorable conditions on Earth that support life.

So much for the "no," but now to the "yes." Fine-tuning, the Big Bang, and stars and planets have a property that I would like to compare to religious icons. Icons (Greek *eikon*, for "image," as opposed to Greek *eidolon*, "illusion") are sacred images that are traditional objects of veneration in Eastern Orthodox churches. They portray saints and biblical scenes. Icons are affixed to the iconostasis, the wall separating the congregation from the sanctuary in liturgical settings, but are also displayed in the prayer corner of homes. Their purpose is not only or even primarily that of artistic decoration. They are instead meant to establish a deep personal and spiritual connection between the beholder and the visualized potential of the scene depicted. Ideally, the depicted person or event takes on a vibrant presence, such that beholders are drawn to respond to it and to participate inwardly in the transcendent element of what is represented. The icon aims at mediating an emotional and existential relationship to the saint or scene and, beyond that, to the Divine. An icon serves as a crevice through which we may glimpse a deeper reality.

God cannot be perceived directly in the fine-tuning, in the Big Bang, or in the formation of stars and planets. However, these remarkable coincidences and processes whereby novel things form may be regarded as icons. They can address the beholder. If he or she permits, these processes can become transparent so as to reveal divine power, wisdom, and goodness. There are no genuine sacred processes or objects in the universe that are wholly distinct from secular things. Though we cannot suppose that some sort of holy intervention is required for a star to form, it is possible to perceive in such creation the realization of a divine idea. This postmythical perception of the sacred[5] does not make star formation into something removed from natural processes. It remains an icon, as many other things may become iconic and remind us of the transcendent foundation of reality.

10.3 Creation Today

It may be understandable but is short-sighted and finally mistaken to interpret the Big Bang as a direct action of God. It would be like confusing the painted Mary of an icon with the real person who lived two thousand years ago. We must ponder further in order to discuss creation in a theological way. One school of thought in modern theology defines creation as "rational origin" (see page 129) of all things. This theology supposes that God has indeed incorporated a creative principle into the universe.[6] God creates the idea and the possibilities, and thus engenders the truly new. Like a discovery that cannot be revoked, this truly new cannot vanish. It is created forever.

The Swiss theologian Hans Weder[7] has pointed out that new entities in the universe like new stars can become a metaphor for the theological notion of creation. In other words, creation is variously epitomized by the formation of a new star, a new human being, or the universe. When a new thing comes into being, it reflects the distinctive and universal act of creation that has occurred for billions of years, is occurring today, and will continue in the future. Seeing a new star as a present-day metaphor for creation illustrates the perception of creation in today's cosmos. But a metaphor works both ways; it also expands our vision of the formation process beyond the isolated case. The new star is thus recognized as a part of the sustained, all-encompassing process of cosmic creation and acquires meaning.

The concept of creation as continuing in the present (in theology known as *creatio continua*) differs from the notion of a once-and-for-all creation restricted to a mythical, primordial era. It is the effective contrary of deistic theology, where God only plays the role of a watchmaker who, after having done his work, simply sits back and enjoys the ticking clock. Continuous creation includes the idea that genuinely new things form today and will in the future. It affirms as well the principle that novel things may appear spontaneously and thus not according to a plan. And while the idea of continuous creation has longstanding roots in the ancient world, as demonstrated by the Gospel of John, the results of modern science, in disciplines ranging from astrophysics to biology, lend concreteness to this conception. Scientists expecting a revolution in theology will be disappointed. The thesis of Karl Barth, "that science cannot provide succor"[8] to the understanding of creation, still holds as a basic principle. Nevertheless, scientific findings doubtless contribute to the way the concept of creation must be addressed today. To be comprehensible in our time, discourse on creation must be based on a sense of both wonder and fright, born from recognition of the role that chance and necessity play in life and in the world today.

If God is the topic in a scientific context, one may easily get the impression that the two planes of perception—theological and scientific—have simply been conflated. When speaking about "God," however, the language must remain on the plane of participatory perception. The same is true when using theological language to write about cosmic phenomena such as star formation and the Big Bang. These processes are usually associated with science, but they do not have to be, as they can also be the objects of our admiration. The universe and its objects can be addressed on both planes, but never God.

Figure 21: A huge molecular cloud is situated in the constellation Cygnus. The infrared picture at 21 micrometers wavelength shows the heat radiation of dust. Above the middle at six thousand light-years distance is DR 21, where a massive star is forming. It is accompanied by many point-like cloud kernels, in which smaller stars of the mass of the Sun form. (Photo: A. Marston [ESA], NASA)

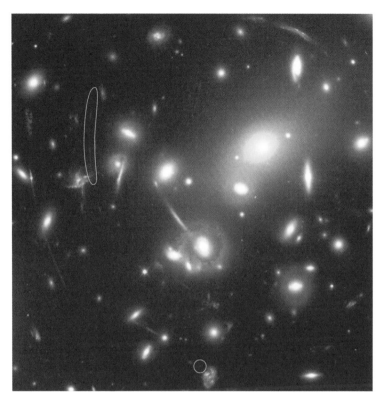

Figure 22: The cluster of galaxies Abell 2218 acts like a lens and enlarges galaxies behind. However, the images of far-away galaxies are strongly distorted. The more distant a galaxy, the more its light is shifted to longer wavelengths. The most distant object ever observed is found in the two contoured areas in the image. Its light was emitted 13.1 billion years ago, just 700 million years after the Big Bang and shortly after the age of the first giant stars. (Photo: J.-P. Kneib, R. Ellis, et al., ESA, NASA)

11

Longing for Meaning

Every now and then, after delivering a public presentation on astrophysics, I am asked the rather strange question: "Did the Big Bang really happen?" The question is puzzling to me since the Big Bang model is one of the most successful theories of astrophysics. Such a question challenges not only our ideas concerning the beginning of the Big Bang, the zero point of time, but the whole scenario of cosmic expansion out of a dense and hot gas until the formation of the first stars. Yet the Big Bang model can explain many observations touching on issues such as the present expansion of the universe, microwave background radiation, the formation of galaxies, and the cosmic abundance of helium. The theory established itself only against great resistance by other scientists and was extremely carefully tested. Certainly, every theory could turn out to be false, but useful theories—those able to explain much in one formulation, those that have been tried and tested for half a century—rarely fail. Technical equipment is designed and constructed in accord with such theories. By applying them in our daily lives, we accept dozens of scientific theories every day without question. I suspect that those asking about the Big Bang are worried not so much about strictly empirical reliability as about a much deeper, more fundamental question: Does the scientific model of the Big Bang threaten to empty the universe of its last vestige of spirituality? If there were a physical explanation of its origin, would the universe become a meaningless mechanism driven by blind chance and impersonal laws? Actually, those posing the question want to know nothing less than whether the universe has a meaning.

The word "meaning" may be viewed in several ways. It is related to the verb "to mean" and "to be meant for." In the workshop of a craftsman each tool has its meaning: a particular place and an assignment for use.[1] A hammer may be used to drive a nail into something; this is its direct purpose. Yet it may also be combined with tongs, for instance, to bend a piece of metal, with a chisel or a screwdriver. The mutual combinations and dependencies enable the proper functioning of the whole workshop to give meaning to any particular tool. Meaning cannot be confined to the object in isolation, but derives from the object's being part of a framework of sense and its contextualized relation to a larger whole.

11.1 Big Bang or Creation: What Is the Question?

I have noticed disappointment on the faces of persons who have asked me if the Big Bang is true, only to receive an answer based on physics. The beginning of the universe apparently has a spiritual importance in the life of those asking such questions. In the view of physics, however, the formation of a new thing, even of the universe, is a natural process. It motivates astrophysicists to describe it with ever-improved mathematical models. Scientific progress takes place on another plane from spiritual inquiry and experience. Its success may overshadow spiritual experiences but does not negate or disprove them. Walt Whitman, were he able to hear a public lecture on contemporary astrophysics, could still be enchanted today by gazing on the night sky.

It is not the task of physics to drive spirituality from the lives of people. Astrophysicists should take such disappointment seriously because it thwarts the curiosity that motivates others to pursue or to support research. It is clearly not appropriate to claim in provocative fashion that science has found the universe to be without meaning. The notion of "meaning" does not exist in science; it cannot be measured or put into an equation. Science cannot provide the necessary context for answering such ultimate questions and is thus unable to answer the question of meaning either positively or negatively. The concept of meaning disappeared from scientific language some time ago, once perception was reduced to measurable facts and interpretations limited to mathematical laws and pure chance. This narrowing of attention does not necessarily negate the notion of meaning, however.

Scientific explanation does not necessarily imply a denial of meaning. When the function of a clockwork gear is explained, for example, the gear scarcely loses its meaning so long as one recognizes the purpose and operation of the clock as a whole. It takes more than physics to exclude the possibility of its meaning. My disappointed interrogators probably supposed that the Big Bang scenario implies a mechanistic wheelwork script that must be enacted as a completely deterministic course of events. The real disappointment sets in only if they proceed to interpret all of reality in this mechanical way, ignoring all other perceptions. They forget that reality is larger than what can be measured by science. Only when the strictly mechanistic and materialistic interpretation is totalized, in a way that far exceeds the available scientific evidence, does it become a construed interpretation. This mechanistic interpretation may already be countered on the plane of physics, as the development of the universe is not mechanically determined but remains open in large measure because its future is not certain. Already at the time of the Big Bang, the evolutionary future remained as open as next year's weather on Earth. The perennial question posed about the Big Bang demonstrates how important it is to distinguish between a scientific explanation and a metaphoric interpretation of the whole universe as mechanical clockwork.

One should also recall that in all astrophysical theories a large but unknown fraction of reality lies beyond our observations and explanations. In this regard, inquiry about the Big Bang is no different from that involved in the formation of a star. The physics of extremely high energies is not known well enough to understand the zero point of time. Thus, it is unlikely that we will ever see either the Big Bang or star formation fully explained.

The Big Bang and creation are not conceptually antithetical, but they answer different questions. Someone who asks what the methods of physics can tell us about the development of the early universe will be drawn to affirm theories of the Big Bang. But someone who answers with a formulation centered in creation is responding to questions about the meaning of the universe, within a broader context of inquiry about all of reality. These divergent responses, which originate from different fields of experience, will next be considered in greater detail.

11.2 A Theory of Everything

There is an endeavor in physics that seems to envision the entire universe, including human existence, as encompassed within a completely mechanical course of events. It is known as the "World Formula" or "Theory of Everything." The theory aims at unifying our understanding of all physical forces, although the name appears to define all of reality. An initial success in this realm fell to James Clerk Maxwell (1831–1879), who developed electromagnetic theory. He was able to reconcile the apparently separate phenomena of electric and magnetic forces. Ever since it has been an important project of theoretical physics to unify the other forces in an all-embracing theory. Thus far, this project has succeeded in incorporating the weak interaction (a force in the atomic nucleus) and the electromagnetic force. Yet the Grand Unified Theory, which would incorporate as well the strong interaction in the atomic nucleus, remains elusive. Today's theoretical approaches expect that the three forces are equally strong at very high energies and are aspects of a single force. Current attempts to complete the Grand Unified Theory suggest that protons, the building blocks of all atomic nuclei, are not stable and will decay. Yet experimental evidence is still missing. The search for proton decay is one of the few tests of the Grand Unified Theory that does not require an effort exceeding the abilities of laboratories on Earth. The elementary particles that must be observed to test the theory directly have an enormous energy that occurs only in cosmic processes.

For completeness, gravity would also have to be unified with the other forces. Such theories are known as "Theories of Everything." There are currently at least four different candidates. All of them are still very hypothetical. A theory that unifies all forces would be necessary to understand the first fraction of a second of the universe. The current theory of gravitation, general relativity, cannot describe those conditions since gravity was so strong that space was warped on the smallest of scales. The curvature of space interfered with quantum mechanical uncertainty, and vice versa. Thus, the first instants of the universe cannot be understood scientifically without reference to a theory capable of recognizing complete unification of all forces.

The names "World Formula" and "Theory of Everything" are definitely misnomers. Intended originally as a joke,[2] "everything" meant to designate the four forces only. These names without context promise too much. Even if the unification project could finally be completed, an infinite remainder would be left to research in physics alone. Fundamental equations cannot explain the immense number of phenomena found in reality. For example, how could these equations explain how galaxies, stars, planets, human brains, and pieces of art can form from elementary particles? A Theory of Everything, if ever found, could only be a theory for the particle-physics aspect of reality. The title of such a theory is not only wrong but presumptuous and can easily be misunderstood (and often is) to suppose that reality is ultimately founded on particle physics. Thus we again encounter a case where all of reality is reduced without saying to the scientifically ascertainable—in particular, that which is physically measurable.

Admittedly, though, unification is a legitimate goal of physics. The most important goal of theory is to describe a multitude of measurements with a single equation. The more different phenomena a theory can explain, the more important it is. A well-known example is Newton's theory of gravitation. It can explain why an apple falls to the ground but also provides the reason why Earth orbits the Sun. Unification can pertain only to the part of reality measured and observed by physics; and physics must be aware of this limit. The discussion of the Gardener Parable (page 63) and of Pascal's vision of God (page 98) illustrates how reality exceeds the range of physics. This point applies all the more to the range of particle physics.

One simply cannot entertain the idea of a totalizing formula for star formation. The formation is a chaotic process that occurs in unpredictable places within the hodgepodge of a molecular cloud. Sure, the chaotic processes are based on very few basic physical equations, but in the end stars and planetary systems are scarcely all alike. Laws and chance are inseparably commingled. The universe is neither a machine nor a computer, and the analogy to the theory explaining a mechanical clockwork is far from adequate.

11.3 Is the Universe a Unity?

Science, and physics in particular, reduces the processes operating in the universe to elementary subprocesses that may then be expressed in mathematical equations. The view of the universe as a whole may get lost in the process. Each single component, like the above-mentioned gear of a clock, can have meaning only as part of the whole. But is the universe a whole, and if so in which way?

The fundamental equations discussed in the previous section already indicate a cosmic unity. Four hundred years after Galileo Galilei, most people have ceased marveling that nature follows mathematical rules. Eugene P. Wigner attributes this attitude to "the unreasonable effectiveness of mathematics in the natural sciences."[3] It is simply astonishing that the physics of the universe is based on a few fundamental equations that hold true everywhere. They are so simple that students can master them in four years. And they yield good physical explanations even for remote galaxies. There are no indications that laws holding in other regions of the universe are any different than those operating in laboratories on Earth.

The constants of physics appear to have the same characteristic; they are the same throughout the universe. One of the constants, the fine-structure constant having the value of 0.0072973525, describes how the spectral lines of hydrogen split into fine structures. The constant has no dimensions such as miles or seconds. It is a pure number. If extraterrestrial civilizations with different units measured it, they would arrive at the same value. It combines several elementary constants including the speed of light, electron mass, and Planck's constant. If one of these constants were to change, so would the fine-structure constant. It is easily measurable and plays a major role for historical reasons. Currently there is a debate as to whether the fine-structure constant has changed, as claimed by a reputable research group, by half a billionth in the past 2 billion years. The constant can also be measured in remote galaxies that can be observed in an ancient state because of the finite speed of light. There the constant was confirmed to be the same, within a millionth of today's value, ten billion years ago.[4]

Since the fine-structure constant combines several important constants, many physical processes would react significantly to any variation in it.

Uniform laws and constants indicate an inner coherence and a genuine unity of the universe. The few building blocks we know to exist are the same throughout the entire universe. There is no explanation for this and no directly apparent plan; it seems to be a property of the universe. We are also part of this surprisingly uniform whole. The elementary particles in our body follow the same cosmic laws and constants as in the remotest stars. We are thus embedded in the unity of the universe.

11.4 Spooky Actions at a Distance

The universe's unity may reach deeper as a consequence of quantum mechanical entanglement. It is the craziest property of quantum theory. In 1935, Albert Einstein and two collaborators pointed out what they called a "spooky action at a distance" to demonstrate how absurd quantum theory is. The theory describes the motion of a particle as a probability wave. According to the customary Copenhagen interpretation, the particle becomes manifest only when it is observed. Before measurement, it is not determined where the particle is or how it moves. Location and speed are uncertain, according to Werner Heisenberg's principle. The values that are finally measured are arbitrary within a certain range. To be precise, location and speed are complementary in quantum mechanics: if one measures the location exactly, one loses all information about speed, or vice versa.

Entanglement now limits the methodical reduction inherent in physics. It is a most surprising and counterintuitive property that manifests itself in the following way: when two particles meet and there is a force at work between them, quantum theory describes them as a single system. The same is true when a particle decays into two smaller particles; the original particle is a quantum mechanical system even after its decay. If the location of one daughter particle is determined, its velocity becomes accordingly uncertain. Now the same holds for the other daughter particle, too, because the two particles constitute an entangled whole that extends over the spatial separation. If one portion of this unity is measured, it immediately affects the comprehensive whole and is not limited by the speed of light. The particles cannot be regarded as two completely separated points in space. Quantum entanglement applies regardless how large the distance is between the particles.

Is the whole universe a quantum mechanically entangled unity? In principle, entanglement applies for all systems with two or more particles. Its effect has been demonstrated in macroscopic objects like salt crystals, and is claimed in photosynthesis.[5] Virtual particle pairs, electrons, and positrons that come spontaneously into existence in a vacuum and subsequently disappear are entangled. Each photon, as the particles of light are known, forms in a quantum mechanical process. The light of distant stars in a stellar atmosphere consists of a large number of such particles if measured by a photon detector. Each photon emitted by an atom in the stellar atmosphere remains entangled with the atom, so long as it does not forget its past by interacting with other particles on its way to us, and if the atom does not collide with another atom in the stellar atmosphere. When such an admittedly rather hypothetical photon struck the retina of our eye, it would cause an instantaneous change in the atom of the stellar atmosphere where the photon originated.

Quantum entanglement does not open the floodgates to cosmic holism, or postulate that everything is entangled with everything and that mutual relations constitute universal reality and unity. The entanglement effects are probably weak in general and have little influence on the scale of heavenly bodies. However, it is not clear yet what happens when many particles are entangled. Quantum mechanical entanglement is a dark unknown in the universe. It can serve to remind us that our current human knowledge is an island in a sea of ignorance.

11.5 Is Humankind the Goal?

One attempt to assign meaning to the universe takes the form of positing a tangible goal to evolutionary processes. The Anthropic Principle is presented here as an example. The principle has been formulated in many versions. In its "strongest" version, it postulates that the emergence of humankind can be viewed as the inevitable endpoint of developmental history. Is the meaning of the universe to form humans?

The Anthropic Principle originated in a much weaker form. In the early 1970s some astrophysicists asked why certain constants in the universe possessed just the right values that enabled life to form. For example, if the fine-structure constant were different by 4 percent, no carbon would be produced in the interior of stars. The whole universe would be altered, and we would not be here. Why has the constant just the right value? For a physicist, a constant without explanation is troubling. It is like the boss's setting an employee's salary without any reason or justification: "This is your pay, period!" There are 19 constants in the Standard Model of particles that cannot be derived; they are unexplained and have to be measured. Why is the universe so finely tuned that life is possible?

Brandon Carter demonstrated through the *weak* Anthropic Principle that in this case a causal explanation is not required: "What we may expect to observe must be restricted by the conditions necessary for our presence as observers."[6] In other words, humankind is part of a cosmic development that unfolds according to certain laws and constants. If they were not as they are, there would be no human beings. We cannot expect to observe what contradicts our existence. If the fine-structure constant were different, we would not be here and could not measure it.

There is also a *stronger* version of the Anthropic Principle. It states that the universe is constructed so that the human species must inevitably emerge, that humankind is a cosmic necessity, and the goal of the universe is to bear humans.[7] The universe would then be like a factory specially built for the production of certain products, including humankind. This version of the principle appears to be an anthropocentric overstatement. It disregards the randomness of biological evolution and the small scale of humanity in relation to the immensity of cosmic developments. Fine-tuning cannot be taken as proof of an inevitable goal, and thus cannot in itself provide meaning. But cosmic fine-tuning remains unexplained and is simply amazing. It can, as noted before, serve an iconic role toward the discovery of meaning if we permit it to do so.

11.6 The Question of Meaning

Anyone looking up into the night sky will eventually wonder about the meaning of the cosmos, and the question, "What does the universe mean?" can certainly ignite vehement discussion about matters such as evolution, the Big Bang, and Intelligent Design. But perhaps these questions themselves only assume their true gravity if we go one step further and ask, "What is the meaning of my existence?" So let us pursue first that primal question about the meaning of life.

In human existence, we experience meaning only in relation to a larger context. People find themselves situated within some larger, more encompassing frame of reference. It is a typically modern question that became more insistent the more persons came to think of themselves as self-reliant individuals. Three kinds of existential experience potentially revelatory of the "meaning of life" might be identified here: (1) People may find meaning in work that enriches

their lives and that of the society they inhabit; or in perform-
ing certain duties, so as to gain the respect of their community; or in diverse forms of social engagement or interaction.
In these cases, persons experience a meaning derived from
their own activities but directed toward fulfilling some
social purpose. (2) Meaning may be experienced as the fruit
of personal relations, whether discovered in the love of fel-
low human beings or in the love of God. Love makes life
meaningful by situating one's existence within the dynamic
context of giving and receiving. (3) A person may experience meaning within the process and consciousness of following a life plan. Those who can identify a divinely shaped
vocation for their lives may hope to set their personal histories into a larger, meaningful context.

There are close connections between these existential patterns of meaning and attempts to discover meaning in the
universe. The European Middle Ages pictured the world as
a theatrical stage for human life. Humanity was taken to be
the ultimate goal of creation, so that the universe's meaning came to be defined rather narrowly as the stage upon
which the human drama was enacted. Today we perceive
the immediate stage of human habitation to be infinitesimal
in relation to the universe as a whole. Thus, the stage metaphor seems embarrassingly anthropocentric today. Nevertheless, this medieval view of cosmic meaning persists
today in some aspects of the strong Anthropic Principle
hypothesis as well as in Intelligent Design. Both provide the
universe with an inherent purpose—that of bringing forth
living creatures—and thus give it meaning. The present-
day Intelligent Design model draws on the traditional stage
metaphor and construes the universe as an instrument
intended to form humans.

Interpreting the universe as a creation gives the universe,
as well as human existence, an ultimate foundation in the
will of the creator. It thus discloses a meaning founded
beyond ourselves and our world. One need not embrace

the idea of a preexistent design to interpret cosmic development as divine creation. On the contrary, if creation by the free will of God happens even today, there cannot have been a design that was fixed billions of years ago. The creation interpretation should nonetheless be understood as more than just another rational construct. In fact, *only a person who is able to apprehend the world emotionally as a gift can interpret it as creation.* The gift touches upon the whole of cosmic history, from the distant past to the unforeseen future. And the notion of creation presumed here includes not only a belief in what is true, but also an existential reliance on the present and an "assurance of things hoped for"[8] in the future. Only insofar as this reliance is confirmed through experience in life can creation solidify to yield the meaning of the universe.

Human beings can thus discover meaning emotionally and experientially, as well as rationally. It happens in relation to a spatial frame of reference vastly larger than the self. Meaning can also be comprehended in relation to temporal contexts reaching well beyond the present moment, in past events of personal, historical, or cosmic significance. Yet the experiential layer of response to the universe most obviously relevant to meaning lies in anticipating the future. According to Christian theology, what is meaningful concurs with salvific history — with "salvation" to be understood not only as human rescue from sin and death but as universal deliverance and wholeness. In this tradition, continuous creation reflects a cosmic striving toward completion and perfection.[9] Temporality then becomes significant, assuming a goal and direction. Time does not simply run in a circle of endless repetition, but aspires toward the new. All that is truly new in the universe, then, being part of a dynamic associated with this goal of completion, becomes meaningful. From such a perspective, the emergence or welfare of humankind is not the sole purpose or goal of the universe. Yet human beings can find meaning in their lives by knowing and experiencing that they are part of this cosmic development.

12

Space and Time— Surprised by Creation

When we gaze at the starry sky with a modern view of the universe, we are inevitably reminded of the smallness of human beings. We are led to realize, too, that the conditions necessary for life to exist seem to be found only in a minute sector of the universe. The temporal limitations of humanity, of all life forms, and of the solar system are no less constricting. What does it mean to be at home in this universe and to have a life's time available?

12.1 The Frightening Size of the Universe

The disputes between geocentric and heliocentric world models in the sixteenth and seventeenth centuries were not only about locating the center of the solar system. The size and the meaning of the cosmos were also in question. In the worldview of the time, the sphere of the fixed stars marked the edge of the universe. The edge was imagined as the surface of a sphere slightly beyond the orbit of Saturn. Astronomers thus speculated the end of the universe to lie at a distance of ten to twenty thousand Earth radii. Since distances in the solar system were underestimated by a factor of ten,[1]

171

the universe in those days would actually have fit between Sun and Earth. A small world!

Some astronomers—even Tycho Brahe (1546–1601), the most eminent observer of that time—were opposed to the heliocentric world system. There were several reasons, including an astronomical one: if Earth encircled the Sun, the stars—if not at infinite distance—are seen under a variable line of sight. The apparent position of a star seen from Earth should trace a small ellipse in the sky, known as yearly parallax, during the course of a year. The instruments of the time did not observe this effect. Either Earth did not orbit around the Sun, or the stars had to be more than 700 times farther away than Saturn. The second option was an abysmal thought for Brahe. How could God have made the universe so large if only a small fraction of it was necessary? What meaning could this huge space have? Around 1660, when the heliocentric model became more and more established among scientists, Blaise Pascal expressed his reaction: "The eternal silence of these infinite spaces fills me with dread."[2] Growing knowledge no longer heightened the sense of humanity's greatness. It rather became a hint of its littleness. If most of space were without meaning, the meaning of the whole cosmos and even of humanity seemed questionable.

When Friedrich Wilhelm Bessel measured the first stellar parallax in 1838, the distance to the nearest stars turned out to be not 700 times but about a million times farther away than Saturn. The nearest star, Proxima Centauri, is in reality at a distance of about six billion Earth radii. If Earth were as large as a pinhead, Proxima Centauri would be as large as a cherry at a distance of 2,500 miles. The size of the observable universe jumped by a further factor of a million in the 1920s when the distance to the Andromeda Galaxy was established. The gas from which the cosmic background radiation originated (detected in 1965) is known to be another factor of ten thousand times farther away. The observable universe is a hundred million billion (10^{17}) times larger today than modeled by Tycho Brahe four hundred years ago.

At the opposite end of the size spectrum, a similar chasm appears: The smallest parts of an atomic nucleus that have a measurable size, the neutrons, are as much smaller than Earth as the Earth is compared to the known universe. Space between atomic nuclei appears to be as empty as space between the stars. Both horizons of the cosmic large and the nuclear small lie far beyond the dimensions of our daily life. A near horizon may cause a feeling of constriction in mountain dwellers, but also of security. One may feel lost surrounded by intangibly distant horizons. In cosmic dimensions, we find ourselves on an incomprehensibly large plain without discernible limits.

Today, however, space is known to be neither empty nor quiet. Even between galaxies there is one hydrogen atom or proton on average in four hundred cubic feet. In addition there is an abundance of neutrinos. These are extremely light-weight particles that originated in the Big Bang and in supernovae. They traverse the universe with almost no interactions. Neutrinos from the early universe have not yet been detected. Their density is estimated to be several thousand per cubic inch. An even larger number of dark matter particles is predicted in this same volume. They, too, are not yet detected but reveal their presence through their gravitational mass, which amounts to five times the mass of ordinary matter in the universe.

Physics has established that there are no genuinely empty spaces. Even a vacuum is not a nothing without properties, since the laws of physics hold in every space. It is perhaps most surprising that energy does not vanish in a vacuum. In the current view of the universe, the vacuum energy exceeds the total energy of matter, regular and dark, by a factor of three. The dark energy of the vacuum accelerates the expansion of the universe. The vacuum is full of energy. Its amount is not constant, but fluctuates according to the

laws of quantum mechanics. Fluctuations mean that particles are created and disappear again locally. These particles
are said to be virtual. They come in pairs, because their electrical charge and other quantum numbers must add up to
zero. Particle pairs appear like whitecaps on ocean waves
and disappear without a trace. Space beyond Saturn is not
empty, as Brahe and Pascal envisaged. Particles and waves,
energy and physical laws fill the space between stars with
a creative dynamic that is a momentous part of the universe. All these particles and energies are essential for the
development of the galactic environment where conditions
in some little corner are suitable for our existence. They do
have a relation to us.

12.2 Empirical Dimension of Time

Time's horizon has also expanded enormously. Newton
and his contemporaries assumed a cosmic age of 6,000
years. Today the horizon in time is at the Big Bang, more
than two million times that age. Cosmic changes within a
human lifetime are minute; yet they are potentiated over
billions of years. The age of the universe allows for a development that was simply inconceivable by our forefathers.

Even the concept of time has changed, more significantly
than the simple age of the cosmos. For Immanuel Kant, time
was an a priori necessity for perceiving reality. Yet time is
more than a condition that our senses require to structure
data. In the past century, physics has discovered some surprising empirical characteristics of time. In relativity theory,
time depends on the frame of reference. Time passes more
slowly for an observer who moves relative to us. A strong
field of gravitation also changes the course of time. Time in
the uppermost floor of a building on earth passes slightly
faster than on the ground floor where gravity is higher.
Thus, to some degree, time can be manipulated by humans.

In the first half of the twentieth century, the dominant

opinion in physics was that time was in principle reversible. To nonphysicists this may appear to be an ivory-tower view. It has its roots in the surprising property that all fundamental equations in physics are reversible in time. So time could take a negative direction and run backward without causing a physical or mathematical inconsistency. The evident direction of time in everyday reality would then only be apparent, and the direction of time would result only from certain processes that are more probable in one direction than in the other. Thus, for example, a tree may fall. It has never occurred, however, that a fallen tree straightened up. According to the fundamental equations, this should be possible since the energy of the fall is conserved and still available in the atoms of the tree and the ground. The more probable path is the one where entropy increases, as the second law of thermodynamics puts it, and the energy dissipates into disordered atomic motions, heating the environment. Does probability define the course of time?

In the 1970s Ilya Prigogine[3] and others stressed the irreversibility of time and claimed it to be a fundamental property, despite the apparent reversibility of the equations of physics. Their rationale was quantum mechanics, the basis of physics, according to which time becomes irreversible through the act of measurement. Reality forms only through the action of observing it. The transition from not-knowing to knowing is irreversible. Therefore, the world of quanta, and thus the whole universe, becomes real only by irreversible processes. Reality is only available at the price of irreversibility of time.

The nature of time became even more intriguing when astrophysics realized that time is not boundless. It apparently started with the Big Bang. This conclusion is consistent with all observations of the distant universe over the last eighty years. It is theoretically possible to avoid positing a beginning of time, through a mathematical trick that introduces an imaginary component of time, in addition to real time.[4] The beginning would be like the North Pole, at which it is not possible to go farther north; there is simply no point farther north than the pole. But the oddity remains: that space and time, as defined in physics by actual measurement, did not exist before the Big Bang and came into being with the Big Bang.

Following latter-day discoveries concerning the transitory nature of cosmic objects, our perspective on time has shifted again today from what it was in the 1970s. It is now recognized more clearly than ever that in a universe with a definite beginning, and in which all things at some point form and subsequently decay, time is neither a dateless continuum nor a matter of course. Everything has its time. Not only the life we know as individuals but our species, *Homo sapiens*, has temporal limits and may well become extinct. Sun, Moon, stars, galaxies, and perhaps all matter in the universe will eventually decay. The inevitable alterations of formation and decay must be recognized as integral to our perception of time. In the world of real objects, time without beginning and end does not exist. Ever since the Big Bang, time has been duration.

12.3 The Origin of Time

There is no comprehensive law of nature governing the formation and existence of time. And yet, without time there would be no cosmic development and thus no universe. Time is an a priori necessity for human perception, for without evolution we would not be here and would have no perceptions. Moreover, time is also an a priori property of the universe. Without time, the universe would not exist and develop. Beyond these fundamental observations, science cannot go.

Is every second a creation from nothing? The question exceeds the range of science, which searches for the cause of a phenomenon based on some law or chance. The progression from cause to effect requires a time before and a time after. Chance, when throwing dice for example, also occurs in time and is inconceivable without time before and after. Explanations based on causality or chance assume a continuous flow of time. Although the existence of time is assumed in science, the ground of this assumption cannot be discussed within the framework of science.

Beyond this frame, and in view of time's immense creative and destructive influence, we may nonetheless ask whether the formation of time suggests a divine creativity.[5] There is nothing—not even in the frame of science—that stands in the way of our interpreting every second as a new creation. Still, the absence of any causal explanation for time's existence does not in itself require a divine creation. Even the beginning of time in the Big Bang does not compel science to invoke the notion of God. That something forms, rather than nothing, offers no compelling evidence for a supernatural force. In science, the existence of time is seen as the regular state of things that does not have to be explained.

How, then, might time be recognized as creation? Hans Weder notes that we encounter creation when we discover *with a sense of wonder* that something has been given to us that we could not bring about ourselves but that is essential for our existence.[6] In this experience we become aware that while we cannot support ourselves, we need not be afraid. It is not necessarily the first discovery of something; rather it is perceived in a different way. The sense of wonder is a necessary condition for this privileged perception. It implies that the experience of creation cannot be objectified. The experiencing person cannot be replaced arbitrarily. A sensate human being, conscious of himself, is a necessary part of this perception. He participates in and resonates with the sense of wonder. Moreover, this human being must want to participate in the perception—rather than thoughtlessly bypassing the issue of time, for example, and not recognizing it as an essential medium of creation.

The history of humanity covers only a minute portion of cosmic time, which started 13.8 billion years ago and will continue for billions of years, until such time as *Homo sapiens* may no longer exist. In our galaxy, the Milky Way, about ten stars form and a few decay per year. A myriad of planets, most of them forever uninhabited, form together with their central star. Galaxies with hundreds of billions of stars turn majestically around their axes in some hundred million years and will eventually decay. We are part of this development at an apparently inconspicuous location in the cosmos, 26,000 light-years from the center of the Milky Way. We enter this development as soon as we exist; we play our small part in it and then disappear. Our species has been drawn into the larger development without being asked, in the same way that each individual human is drawn into the perils of earthly history. We are one of 350,000 generations who are living out their time since the genus of *Homo* separated from that of today's chimpanzees. We participate in, but also float along with, a huge river of time.

It may be that, at a certain point in life, we experience a special minute that is more than just a self-evident continuation of the 72.6 million million ($7.26 \cdot 10^{13}$) previous minutes that have passed since the Big Bang. We may experience it as *graciously granted time*. In such a moment we experience time as creation. It may feel like a veil is being lifted from our eyes when we realize that Earth and humanity are not simply thrown into time, and are not here simply as a matter of course, but that the universe and cosmic developments are granted a certain duration. Cognizant of our limited lifetime, we may then also perceive that time is given to us personally. Time is granted not only to us, but to everything else in the cosmos! In this respect we can understand ourselves as on a par with all that is, throughout the universe.

"Time is given to us" is a metaphoric expression for experiencing time as if it were a gift, as something other than self-evident. From this standpoint, the ticking of the clock is no longer experienced as defining the normal state of time. Real gifts are something extraordinary and undeserved. To be sure, we might also describe as "gifts" certain offerings that are in fact either obligatory or generic. But I am speaking here about a gift of grace, such as might be conferred upon a poor farmer given a piece of land by a gracious king, or conferred upon a condemned culprit lacking any claim to receive it by a merciful judge. Sometimes we discover that another minute, day, or year is just given to us undeservedly. The gift of time does not become our property; it is like a fief for a certain time.

In the metaphor of the gift, the giver of time becomes a person. It should not bother us that human traits are commonly attributed to the creator. How else should one talk of the creative and generous force and presence if not in anthropomorphic terms? Moreover, the metaphorical reference to a creator need not imply that the One so named is to be understood in wholly isolated, self-referential terms. Here we are not speaking of a universe to which we have added God. Time and its creative power include all processes in the whole universe in the past, present, and future. The giver of time is part of the great metaphor wherein a gift is bestowed, an exchange in which the receiver also plays an integral role. Humans are those who receive the gift and are capable of perceiving it as such. And it is fitting that the one who receives should be moved to thank the giver. There are some remarkable verses in the psalms[7] where the psalmist praises God even in the midst of abandonment and loneliness. The last words of Jesus on the cross incorporate one of these psalm verses. To praise is the suitable emotional response to receiving something important.

12.4 Outlook

Today we can no longer feel justified in interpreting the universe, in the way that medieval Europeans did, as merely the stage upon which the human drama is enacted. Even less can we suppose the cosmos to constitute merely the stage for our own cameo appearance as mortal individuals. Nonetheless, surpassing the usual limit set by the speed of light, the human mind is capable of sweeping to the edge of the observable universe so as to engage conceptually matters far beyond its earthly sphere. In so doing, we humans have today come to apprehend the universe as not merely the stage, but the play itself. We can now appreciate that this drama includes many forms of enactment in addition to its most familiar scenes of human life and history. Moreover, its conclusion seems to remain open. We cannot easily identify with what earlier generations may have envisioned more assuredly about the ideal completion or goal of cosmic development. The universe, like our lives, does not move in a straight line toward an obvious goal. And we need to recognize the metaphorical limitations, as applied to the cosmos, of the very idea of completion or of a temporal endpoint such as human creators aspire to reach in their artistic works.

The way we interpret our lives is the way we interpret the universe. Interior and exterior, beginning and end, are connected in a closed but enormous loop. Accordingly, our perennial search for meaning can never be satisfied within the narrow bounds of our mortal lives but requires engagement with the larger whole to which we belong. To interpret ourselves and the world as creation means to understand and to enjoy the realization that the essential conditions for life are given to us by a gracious hand.

Figure 23: The protostar AFGL 2591 is just a few times ten thousand years old, yet has a mass sixteen times that of the Sun and is still growing. It is at a distance of about ten thousand light-years. The bright circular area in the center is the collapsing envelope. An outflow of hot gas shoots out to the right. The image was taken in infrared light. AFGL 2591 is deeply imbedded in its natal cloud core and not visible in optical light. (Photo: C. Aspin et al., Gemini Obs., NSF)

Epilogue

The James Clerk Maxwell Telescope is located on Mauna Kea, the highest mountain in Hawaii, 12,000 feet above the breakers of the Pacific Ocean. The diameter of the dish-like mirror is fifty feet. It collects from the cosmos radiation that has a wavelength of a millimeter or less. An operator controls the instrument. I sit with him in a little room that moves along with the telescope, following the stars. I am the observer and follow the measurements on the screen. We observe only at night, when the humidity is low and the Earth's atmosphere is more transparent to submillimeter waves.

We try to substantiate that the protostar AFGL 2591 emits X-ray radiation, which is absorbed completely by the envelope. In this process special molecules are produced, the rotational emission of which we try to detect with the telescope. The radiation's wavelength, between infrared and microwaves, is large enough to escape the dust and gas of the envelope, delivering messages to us about the interior.

The observers' room feels confining, and I find my way to the outside. On the mountain no artificial lights are permitted outdoors, as they would disturb other telescopes observing optical light. As I move away from the telescope, the darkness descends over my eyes like a veil. Slowly they adapt, and a sea of billions of stars illuminates my way. The moon is not up. The brilliance of the unfamiliar starry sky is overwhelming. I look for the constellation of Cygnus, where the telescope is pointing, and find myself surprised that the position in the sky is disappointingly empty and unimpressive. Of course AFGL 2591, which we are currently observing, cannot be seen. It is just forming and is still surrounded by a thick envelope that is falling onto the central object. The envelope is opaque to visual light that human eyes can detect. The dark molecular cloud is about half a finger wide by the measure of the outstretched arm. It contains dust that completely absorbs light. Dust is the material from which planets like Earth may form and contains all elements that are necessary for life.

It is chilly. I am cold despite my winter jacket. I walk leisurely up the road toward the Subaru telescope. We always have a goal for our daily activities and consciously or not ascribe some meaning to them. Thus we accept a certain meaningful order in our little sphere of activity. My immediate goal is clear: the summit ridge of Mauna Kea.

The little dark cloud in Cygnus is now exactly above the crater cone of the extinct volcano. The cloud is about ten thousand light-years away. I try to imagine its geometry. It is known from previous observations in the infrared light range that two powerful outflows punch through the collapsing envelope. What drives these outflows? Again, nothing can be seen of them by the naked eye. For every answer, new questions arise. It appears that we know less and less, the more we observe. It suddenly occurs to me that the scientific method may thus end up frustrating our ambitions and never reach an end. Instead of leading us to full clarity of knowledge and understanding, as we have come to expect since the Age of Enlightenment, science may finally reveal more of our own ignorance. The complexity of star and planet formation may be so enormous that we will never decipher it completely.

AFGL 2591 is not yet a hundred thousand years old. The star will become much bigger than the Sun. In its center, 16 times the mass of the Sun has already accreted. Another 42 solar masses have been measured in the envelope. The details about this celestial body are obviously not of general interest, but they give me a firm base in reality. They help to order my questions on the really important things. Does the formation of a star like AFGL 2591 follow a cosmos-wide impersonal mechanical process, in which I am likewise involved? There is no blueprint for stars, and yet they form. AFGL 2591 is born in the chaos of turbulence and by an immense number of processes. Its formation is not predictable, yet every star finds its equilibrium in which it burns hydrogen like all other stars. The protostar that I know like an old acquaintance in terms of its appearance, traits, and history seems to hold a secret that has something to do with the enigmas of my own consciousness and life.

When stars form, we encounter an inscrutable but apparently friendly universe. Obviously, there is no static harmony. Chaos, formation, and decay are the cornerstones of cosmic evolution, and they frighten me. In some moments, however, an affectionate face seems to shimmer through it all, even through decay. It disappears as soon I want to focus on it. Is not the complexity of life at least as inscrutable as the universe? Also in life, there are moments where goodness is experienced. Psalmists nearly three thousand years ago praised both, life and universe, and today praise could still be an appropriate response.

The wind becomes stronger on the ridge where the optical telescopes are arranged. Telescope motors hum from the Keck Observatory. I turn. Walking downhill, my eyes are attracted again to the dark cloud in Cygnus. It remains mysterious, but the thought pleases me that we have much in common. The sentiment might be called empathy, based on a mutual kinship that I cannot fathom with telescopes. The appeal of this mysterious recognition will not divert me from my work, however. On the contrary, scientific findings appear to me like marvelous flowers, strewn along a path that I take in consort with the whole universe. Our mutual destination is unknown. Yet I feel carried along during my nightly walk, and with the benefit of this emotional perception may well interpret the whole universe as being carried along as well. To be supported even in the face of dire catastrophe is the experience I associate most closely with the presence of God.

How can astronomy still fascinate me after all these studies and years of professional research? When I look up to the starry sky on a clear night as a professional astronomer, it seems to me that the stars glint more brilliantly than they are obliged to do according to black-body theory.[8] The deeply enigmatic sense of things I experienced earlier, that one night out in the Sahara, has remained—precisely because the creator remains invisible, and because I find his relation to our future and the future of the cosmos as elusive as ever. The questions persist. Yet the direct, experiential perception of reality that stimulates such questions also continues to inspire me. I am no longer amazed by the size of the universe, but I am overwhelmed time and again by the simple fact that I am conscious of my own existence on this flourishing oasis in the midst of a huge, cold, and violent universe.

Acknowledgments

Astronomy unfolds today with a speed that has never been seen before thanks to ever better tools of observation. The specialist literature brims with fascinating news. In theology, on the other hand, new ideas originate in discussions and in disputations. They develop from historical conceptions or in opposition to currently prevailing ones. Thus, dialogue is the driver in theology and also in the interaction with the sciences. I have learned much in discussions with theologians and would like to thank in particular the professors Samuel Vollenweider, Pierre Bühler, Julia Gatta, Michael Nausner, Konrad Schmid, Christopher Bryan, and Jesse J. Thomas. They have helped me find my way through the many fields of their scholarship. Meetings of the Zurich Center of Competence in Hermeneutics and the Ecumenical Working Group on Creation Theology in Solothurn (Switzerland) have contributed many inspirations. My colleagues Roman Brajša, Bruce N. Lundberg, and Ralph Neuhäuser have critically read the manuscript from the astronomical side. The late Elisabeth Benz, Maja Pfaendler, and the late Alfred Ringli have made helpful suggestions to the draft.

A special thank goes to Professor John Gatta, who not only carefully read the English translation but contributed essentially to this book from its beginning.

Notes

Scripture quotations are from *The Holy Bible*, English Standard Version® (ESV®), copyright © 2001 by Crossway, a publishing ministry of Good News Publishers. Used by permission. All rights reserved.

Chapter 1. The Stuff We Are Made Of

1. Isaac Newton wrote in his first letter to Richard Bentley (1692): "The matter on the outside of this space would by its gravity tend towards all the matter on the inside & by consequence fall down to the middle of the whole space & there compose one great spherical mass. But if the matter was evenly diffused through an infinite space, it would never convene into one mass, but some of it convene into one mass & some into another so as to make an infinite number of great masses scattered at great distances from one another throughout all that infinite space. And thus might the Sun and fixt stars be formed supposing the matter was of a lucid nature." English trans. Andrew Motte (New York, 1846), cited from *Letters to Richard Bentley*, in *The Newton Project*, ed. R. Iliffe and S. Mandelbrote (University of Sussex, UK, 2011).

2. Michael Hoskin summarized the historic developments in *The Cambridge History of Astronomy* (Cambridge: Cambridge University Press, 1999), 190-94.

3. Newton's notion of God was rather sophisticated. In his *Philosophiae naturalis principia mathematica* (third book, London, 1687; Engl. trans. Andrew Motte; New York, 1846), 504-6, he says: "And if fixed stars are the centres of other like [planetary] systems, these, being formed by the like wise counsel, must be all subject to the dominion of One; especially since the light of fixed stars is

of the same nature with the light of the sun. . . . This Being governs all things. . . . He is omnipresent not virtually only, but also substantially; for virtue cannot subsist without substance. . . . But by way of allegory, God is said to see, to speak, to laugh, to hate, to desire, to give, to receive, to rejoice, to be angry . . . to work, to build."

4. Winfried Schröder, *Ursprünge des Atheismus: Untersuchungen zur Metaphysik- und Religionskritik des 17. und 18. Jahrhundert* [Origin of Atheism: Investigations on Metaphysics and Critique of Religion in the 17th and 18th Century] (Stuttgart: Fromann-Holzboog, 1998).

5. Black holes are regions in the universe with immense matter density, the gravity of which is sufficient to hold back everything. Even photons cannot escape. Such a region is therefore black if observed from outside. Black holes will be an issue again in Chapter 3.

6. Mark R. Krumholz, Christopher D. Matzner, and Christopher F. McKee, "The Global Evolution of Giant Molecular Clouds. I: Model Formulation and Quasi-Equilibrium Behavior,"*Astrophysical Journal* 653 (2006): 361-82.

Chapter 2. When Stars and Planets Form

1. Angular momentum measures the momentum of the spin of an object. In physics it is defined as the product of velocity, mass, and distance from the rotation axis.

2. Victor Hugo, *Œuvres complètes, Histoire* (Vol. XII), *Choses vues* (Paris: Robert Laffont, 1987), 686.

3. The collision hypothesis was originally proposed in 1749 by Georges Louis Leclec de Buffon, who conjectured that a huge comet initiated the formation of the planets. The collision hypothesis in various versions persisted until the middle of the past century, when Hannes Alfvén noticed the cosmic importance of magnetic fields.

4. Francesca Bacciotti, Thomas P. Ray, Reinhard Mundt, Jochen Eislöffel, and Josef Solf, "Hubble Space Telescope/STIS Spectroscopy of the Optical Outflow from DG Tauri: Indications for Rotation in the Initial Jet Channel," *Astrophysical Journal* 576 (2002): 222-31.

5. Michel Mayor and Didier Queloz, "A Jupiter-Mass Companion to a Solar-Type Star," *Nature* 378 (1995): 355-59. In 1992, two planets that orbit a neutron star were reported by Aleksander Wolszczan and Dale Frail.

6. Pascal Stäuber et al., "Water Destruction by X-rays in Young Stellar Objects," *Astronomy and Astrophysics* 453 (2006): 555-65.

Chapter 3. Boundless?

1. The term "Big Bang" is used in two different connotations: Originally it denotes a model in which the universe started to expand explosively out of a hot, dense state. Today the notion is also used for the hypothetical singularity at time zero, when, mathematically, density and temperature reach infinity. Here we use the term Big Bang in its first connotation. The Big Bang scenario, but not the singularity, is widely accepted by the experts, although some details of today's standard model are still controversial.

2. Richard Feynman, in *The Pleasure of Finding Things Out: The Best Short Works of Richard Feynman*, ed. J. Robbins (Cambridge, MA: Perseus Books, 1999).

3. Walt Whitman, *Leaves of Grass*, 1855 (new edition, New York: Bartleby Company, 1999), 180.

Chapter 4. Origin and Creation

1. Paul Michel, *Physikotheologie: Ursprünge, Leistungen und Niedergang einer Denkform* [Physico-theology: Origin, Achievements, and Anticlimax of a Form of Thought] (Zurich: Beer, 2008).

2. Exodus 3:14.

3. As Georg Christoph Lichtenberg (1742–1799) concluded.

4. Karl Barth, in *The Doctrine of Creation, Church Dogmatics* Vol. III/I (New York: T. & T. Clark, 1986), ii.

5. Frank J. Tipler affirms that religious statements can be proven scientifically in *The Physics of Immortality: Modern Cosmology, God and the Resurrection of the Dead* (New York: Anchor, 1997), 560.

6. According to Artur Weiser, *Die Psalmen* (Göttingen: Vandenhoeck & Ruprecht, 1987), 64.

7. For example in Psalms 19 and 104.

8. Hans Weder (in *Kosmologie und Kreativität* [Leipzig: Evangelische Verlagsanstalt 1999], 66) refers to a word of Jesus for primary experiences of creation: "Look at the birds of the air: they neither sow nor reap . . . and yet your heavenly Father feeds them. . . . Consider the lilies of the field, how they grow: they neither toil nor spin, yet I tell you, even Solomon in all his glory was not arrayed like one of these" (Matthew 6:26).

9. Antony Flew, *New Essays in Philosophical Theology*, ed. Antony Flew and Alasdair MacIntyre (New York: Macmillan, 1955), 96. After an earlier version by John Wisdom, in "Gods," *Proceedings of the Aristotelian Society* 45 (1944): 185-206.

Chapter 5. The Development Continues

1. The general idea of a willed creation may be viewed as an all-embracing, presently and endlessly ongoing design that stands behind all cosmic evolution. It must be distinguished from the modern Intelligent Design movement (Phillip E. Johnson, *Darwin on Trial* [Downer's Grove, IL: InterVarsity, 1993]; Michael Behe, *Darwin's Black Box* [New York: Free Press 1996]), where design is inserted in a few instances of a general mechanistic evolution following pure chance and strict rules. The philosophical background of the Intelligent Design concept is critically discussed by Hans-Dieter Mutschler in "Intelligent Design—spricht die Evolution von Gott?" [Intelligent Design—Does Evolution Refer to God?], *Herder Korrespondenz* 59 (2005): 10, 497-500.

2. Jürgen Moltmann notes regarding the perception of creation: "The knowledge of nature as God's Creation is participating knowledge." See *God in Creation: A New Theology of Creation and the Spirit of God* (San Francisco: Harper & Row 1985), 2.

3. Klaus-Peter Schröder and Robert Connon Smith, "Distant Future of the Sun and Earth Revisited," *Monthly Notices of the Royal Astronomical Society* 386 (2008): 155-63.

4. I.-Juliana Sackman, Arnold I. Boothroyd, and Kathleen E. Cramer, "Our Sun. III. Present and Future," *Astrophysical Journal* 418 (1993): 457-68.

Chapter 6. Living in the Midst of Evolution

1. Leslie W. Looney, John J. Tobin, and Brian D. Fields, "Radio-active Probes of the Supernova-Contaminated Solar Nebula: Evidence That the Sun Was Born in a Cluster," *Astrophysical Journal* 652 (2006): 1755-62.

2. J. Huw Davies. "Did a Mega-Collision Dry Venus' Interior?," *Earth and Planetary Science Letters* 269 (2008): 376-83.

2. David Nesvorný, David Vokrouhlický, William F. Bottke et al., "Express Delivery of Fossil Meteorites from the Inner Asteroid Belt to Sweden," *Icarus* 188 (2007): 400-413.

4. Patrick Michel, Paolo Farinella, and Christiane Froeschlé, "Dynamics of Eros," *Astronomical Journal* 116 (1998): 2023-31.

5. Narciso Benítez, Jesús Maíz-Apellániz, and Matilde Canelles, "Evidence for Nearby Supernova Explosions," *Physical Review Letters* 88 (2002): article no. 081101.

6. Joe Kirschvink, "Snowball Earth," in *The Proterozoic Biosphere,* ed. J. W. Schopf and C. Klein (New York: Cambridge University Press, 1992), 226.

7. A. V. Gurevich and K. P. Zybin, "High Energy Cosmic Ray Particles and the Most Powerful Discharges in Thunderstorm Atmosphere," *Physical Letters A* 329 (2004): 341-47.

8. Rainer Gersonde et al., "New Data on the Late Pliocene Eltanin Impact into the Deep Southern Ocean," *Geophysical Research Abstract* 7 (2005): 2449.

Chapter 7. Reality in the Cosmos and in Life

1. Blaise Pascal, *Memorial*, trans. Emile Caillet and John C. Blankenagel, *Great Shorter Works of Pascal* (Philadelphia: Westminster Press, 1948).

2. Blaise Pascal deliberately uses the name of God given to Moses in Exodus 3:15 following God's self-revelation.

3. For instance, Albert Einstein, in Max Jammer, *Einstein and Religion* (Princeton, NJ: Princeton University Press 1999), 44.

4. One of the seven fundamental laws of the legendary ancient Egyptian sage or god Hermes Trismegistos. The fundamental principle "as above, so below; as below, so above" was important in alchemy and continues to exist in modern esotericism. Mircea

Eliade gave an account in *History of Religious Ideas*, Vol. 3 (Chicago: University of Chicago Press, 1985), 149.

5. John Polkinghorne has explored the analogy between reality in quantum mechanics and theology in *Quantum Physics and Theology: An Unexpected Kinship* (London: SPCK, 2005). He emphasizes experience as the starting point for understanding the world of both science and theology.

Chapter 8. From Perception to Interpretation

1. Paul Ricœur mitigated the sharp separation with the objection that no penetrating comprehension is possible without explanation. For example, in *From Text to Action: Essays in Hermeneutics II*, trans. Kathleen Blamey and John B. Thompson (Evanston, IL: Northwestern University Press, 1991), 161-82. Similarly, Hans Weder, in *Neutestamentliche Hermeneutik* (Zurich: TVZ, 1986), 120, counters the sharp separation with the words: "there cannot be comprehension without the dimension of explanation."

2. For example, Carl Friedrich von Weizsäcker, in *The History of Nature* (Chicago: University of Chicago Press, 1949).

3. Pattern recognition and metaphors in religion are further discussed by Arnold Benz in *The Future of the Universe: Chance, Chaos, God?* (New York: Continuum Publishing Group, 1997), 157-61.

4. For a more complete treatment of the shortcomings of the "mechanical" interpretation of nature, see Jesse J. Thomas, "The Parable of the Chicken House: Charles S. Peirce, Complex/Dynamical Systems, and Environmental Science," *Journal of Environmental Science and Engineering* B 4 (2015): 559-68.

5. Jacques Monod, *Chance and Necessity: An Essay on the Natural Philosophy of Modern Biology* (New York: Vintage Books, 1972).

6. There are several versions of Kurt Gödel's theorem of incompleteness. A well-known version is "Any sufficiently powerful formal system is either inconsistent or incomplete."

7. Ian G. Barbour has discussed "models" of God's role in nature in *Religion and Science: Historical and Temporary Issues* (San Francisco: Harper, 1997), 305ff.

8. Antje Jackelén, "A Critical View of 'Theistic Evolution,'" *Theology and Science* 5 (2007): 151-65.

9. There are basic theological texts legible also for scientifically oriented readers, e.g., Wolfhart Pannenberg, *Systematic Theology*,

Vol. 2 (Grand Rapids: William B. Eerdmans Publishing Company, 2000).

10. R. P. Feynman, R. B. Leighton, and M. Sands, *The Feynman Lectures*, Vol. III (Reading, MA: Addison-Wesley, 1965), 112.

11. Dietrich Korsch, *Dogmatik im Grundriss* [Compendium of Dogmatics] (Tübingen: Mohr Siebeck, 2000), 191-96.

12. Ingolf U. Dalferth writes in *Becoming Present* (Leuven: Peeters, 2006), 120: ". . . if we are told that the concept of God is needed to synthesize the different spheres of our experience, this may tell us something about the structure of our experience but not much about God."

Chapter 9. What Could Creation Mean Today?

1. Pierre Bühler points out that science and faith are an open interactive system in *Science et foi font système* [Science and Faith Are a Coupled System] (Geneva: Labor et Fides, 1992), 74-76.

2. Immanuel Kant distinguishes the rational origin from the origin by time (causality). The rational origin denotes the conditions of the possibility, e.g., of star formation. For the realization of this possibility, a framework of conditions is necessary. See *Die Religion innerhalb der Grenzen der blossen Vernunft* [Religion within the Limits of Reason Alone], Akad.- Ausg. Bd. 6, 39 (Chicago: Open Court, 1960), 34.

3. Fritjof Capra, *The Tao of Physics: An Exploration of the Parallels between Modern Physics and Eastern Mysticism* (Boston: Shambhala Publications, 1975), 224.

4. Hans Weder, *Kosmologie und Kreativität* (Leipzig: Evangelische Verlagsanstalt, 1999), 57.

5. The formation of new entities out of thermal fluctuations was described extensively by Ilya Prigogine in *From Being to Becoming: Time and Complexity in the Physical Sciences* (New York: Freeman, 1980), 124.

6. Stéphane Udry, Xavier Bonfils, Xavier Delfosse et al., "The HARPS Search for Southern Extra-Solar Planets XI. Super-Earths (5 and 8 MEarth) in a 3-Planet System," *Astronomy and Astrophysics* 469 (2007): L43–L47.

7. Anthony D. Barnosky, Nicholas Matzke, Susumu Tomiya

et al., "Has the Earth's Sixth Mass Extinction Already Arrived?" *Nature* 471 (2011): 51-57.

8. Alfred North Whitehead says in *Process and Reality* (New York: Free Press, 1969), 125: "Life is robbery" and generalizes it later to all evolutionary processes. See also Richard Dawkins, *The Selfish Gene* (Oxford: Oxford University Press, 1976).

9. Extending from the well-known ethical maxim "Reverence for Life" by Albert Schweitzer to the dynamics of evolution, including inanimate matter.

10. The quote "Even if I knew that tomorrow the world would go to pieces, I would still plant my apple tree" is attributed to Martin Luther. The first written evidence, however, dates back only to 1944.

11. Jürgen Moltmann points out that hope may not only be a promise for the future, but an "'expectation' . . . which sets about criticizing and transforming the present because it is open towards the universal future of the kingdom [of God]." See *Theology of Hope: On the Ground and the Implication of a Christian Eschatology* (Minneapolis: Fortress Press, 1993), 335.

12. Ina Praetorius points out that the expectation of a positive future yields the resources for ethical action. See *Handeln aus der Fülle* [Acting out of Plentitude] (Gütersloh: Gütersloher Verlagshaus 2005), 93.

Chapter 10. God in the Universe

1. There were also earlier adumbrations of the notion "creation from nothing": e.g., pre-Christian, 2 Macc. 7:28; and later in the second century CE: Shepherd of Hermas, *Visions* 1.1; Tatian, *Address to the Greeks* 5.12; Theophilus of Antioch, *To Autolycus* 1.4.

2. For a fuller discussion of the role of resurrection in modern theology see, e.g., Christopher Bryan, *The Resurrection of the Messiah* (Oxford: Oxford University Press, 2011), 1-41 and throughout.

3. Following Siegfried Schulz, *Das Evangelium nach Johannes* [The Gospel according to St. John] in the series Das Neue Testament Deutsch, 13th ed. (Göttingen: Vandenhoeck & Ruprecht, 1975).

4. Bernhard Haisch notes the ten most critical properties of the universe in *The Purpose-Guided Universe* (Franklin Lakes: New Pages Books, 2010), 207–8.

5. John Gatta, *Making Nature Sacred* (Oxford: Oxford University Press, 2004), 242.

6. Samuel Vollenweider, "Wahrnehmung der Schöpfung im Neuen Testament" [Perception of Creation in the New Testament], *Zeitschrift für Pädagogik und Theologie* 55 (2003): 246-53.

7. Hans Weder, "Widerspiegelung der Kreativität" [Reflectance of Creativity], *Forum theologische Literaturzeitung* 1 (1999): 56-59.

8. Karl Barth, in *The Doctrine of Creation, Church Dogmatics* Vol. III/I (New York: T. & T. Clark, 1986), ii.

Chapter 11. Longing for Meaning

1. The metaphor of the meaning of the tools in a workshop is ascribed to the philosopher Edmund Husserl (1859–1938).

2. The term popped up in the context of string theory, acclaimed by some as a "Theory of Everything" and criticized by others as a "Theory of Nothing."

3. Eugene P. Wigner, "The Unreasonable Effectiveness of Mathematics in the Natural Sciences," *Communications on Pure and Applied Mathematics* 13 (1960): 1-14.

4. Steve K. Lamoreaux, "Washing Up with Hot and Cold Running Neutrons: Tests of Fundamental Physical Laws," *AIP Conference Proceedings* 769 (2005): 674–77. In a different way also Michael T. Murphy, J. K. Webb, and V. V. Flambaum, "Revision of VLT/UVES Constraints on a Varying Fine-Structure Constant," *Monthly Notices of the Royal Astronomical Society* 384 (2008): 1053-62.

5. Mohan Sarova, Akihito Ishizaki, Graham R. Fleming et al., "Quantum Entanglement in Photosynthetic Light Harvesting," *Nature Physics* 6 (2010): 462–67.

6. Brandon Carter, "Large Number Coincidences and the Anthropic Principle in Cosmology," in *Confrontation of Cosmological Theory with Observational Data,* ed. M. S. Longair (Dordrecht: D. Reidel Publishing, 1974): 291-98.

7. The strong Anthropic Principle is described by John D. Barrow and Frank J. Tipler in *The Anthropic Cosmological Principle*

(Oxford: Oxford University Press, 1988). The strong Anthropic Principle was criticized from a theological perspective by Dirk Evers, in *Raum—Zeit—Materie* [Space—Time–Matter] (Tübingen: Mohr Siebeck, 2000).

8. Hebrews 11:1.

9. Jürgen Moltmann emphasized the goal of perfection in creation in *God in Creation* (San Francisco: Harper & Row, 1985), 5. The concept of creation in the future is discussed by many Christian theologians who have been engaged in the dialogue with science, in particular Pierre Teilhard de Chardin, Karl Heim, Wolfhard Pannenberg, and Klaus P. Fischer.

Chapter 12. Space and Time— Surprised by Creation

1. Michael Hoskin, *The Cambridge Concise History of Astronomy* (Cambridge, UK: Cambridge University Press, 1999), 100.

2. Blaise Pascal, *Pensées*, no. 206 (London: Penguin, 1966), 66.

3. Ilya Prigogine, *From Being to Becoming: Time and Complexity in the Physical Sciences* (New York: Freeman, 1980).

4. Stephen W. Hawking, *A Brief History of Time* (New York: Bantam, 1988), 147.

5. Arther Peacocke notes, "God is creating at every moment of the world's existence in and through the perpetually endowed creativity of the very stuff of the world," in "Welcoming the 'Disguised Friend'—Darwinism and Divinity," in *Intelligent Design Creationism and Its Critics*, ed. Robert P. Pennock (Cambridge, MA: MIT Press, 2001), 473.

6. Hans Weder, *Kosmologie und Kreativität* (Leipzig: Evangelische Verlagsanstalt, 1999), 68.

7. For example, Psalm 22:2 and 22:26-32.

8. According to the dictum ascribed to Jürgen Moltmann, "the birds sing more beautifully than they are obliged to do according to Darwin's theory."

Index

About the Author

Arnold Benz received his Ph.D. in astrophysics at Cornell University, NY and was head of the Group for Radioastronomy and Plasma Astrophysics at the Institute for Astronomy of the ETH Zurich, Switzerland. His teaching and research focused on plasma astrophysics, high-energy astrophysics, and the physics of star and planet formation. Arnold Benz was president of Division II (Sun and Heliosphere) of the International Astronomical Union (IAU), president of the Swiss Society for Astrophysics and Astronomy (SSAA), and president of the Community of the European Solar Radioastronomers (CESRA). He was awarded honorary doctorates of the University of the South, Sewanee, TN, in science and of the University of Zurich for theology. He lives in Zurich with his wife and has four children and 13 grandchildren.

About the Publisher

The Crossroad Publishing Company publishes Crossroad and Herder & Herder books. We offer a 200-year global family tradition of books on spiritual living and religious thought. We promote reading as a time-tested discipline for focus and understanding. We help authors shape, clarify, write, and effectively promote their ideas. We select, edit, and distribute books. With our expertise and passion, we provide wholesome spiritual nourishment for heart, mind, and soul through the written word.